Uninhabited Ocean Islands

Jon Fisher

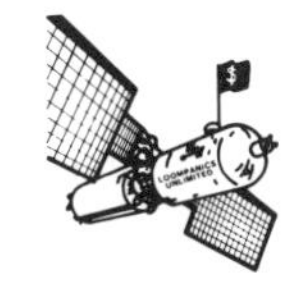

Loompanics Unlimited
Port Townsend, Washington

WARNING: *The maps in this book are for identification purposes only and are not intended for navigation. Neither the author nor the publisher will be held accountable for the use or misuse of the information contained in this book.*

This book was previously sold under the title **Uninhabited And Deserted Islands**. *This is an updated, revised, expanded edition of that book.*

UNINHABITED OCEAN ISLANDS

Published by:

Loompanics Unlimited
PO Box 1197
Port Townsend, WA 98368

Loompanics Unlimited is a division of Loompanics Enterprises, Inc.

Maps by Kevin Martin, Patrick Michael and Jon Fisher

ISBN 1-55950-074-3
Library of Congress Catalog Card Number 91-061945

Contents

Part II — Sub-Antarctic Islands, Pacific Sector

Part III — Atlantic Ocean Islands

Part IV — Indian Ocean Islands

Introduction

You've probably seen cartoons about castaways on deserted islands that are only a few feet across with one palm tree growing in the middle. Well, such uninhabited islands do exist, although they don't look exactly like that. They can be found throughout all the oceans of the world. And the way things are going now, even more islands may be deserted in the future.

Just as people tend to move from rural areas to cities, they also tend to move from islands to the mainland, and from small islands to larger ones. The "bright lights" seem to offer an irresistable lure with their promise of better jobs and a better way of life. The result is that small, unpopulated islands are likely to remain devoid of people, and many other islands where only a small number of people live now may become completely depopulated at some future time.

Besides this rush to urban areas, many empty islands are uninhabited for one of three good reasons: they are barren sandy or rocky places where there is little vegetation; or they are active volcanoes; or they are tiny atolls with very little land area above sea level. These islands range in size from a couple dozen acres up to several square miles.

Barren islands very often support huge populations of sea birds. They have no natural running water, although rainfall might collect in temporary pools. Some barren islands consist of one gigantic rock. A few of these rock islands have been turned into strong fortresses with powerful guns mounted in caves dug deep into the rock. Examples of such forts are the National Chinese island of Quemoy, and the Rock of Gibraltar at the tip of Spain.

Islands that are active volcanoes are usually uninhabitable. Because of the frequent outflows of lava, poisonous gases and ash, very little life can get a foothold in such places. The hellish island may vary greatly over time.

Coral Atolls

Many deserted islands are atolls, which consist of a circular reef built up by coral (whose skeletons accumulate to form a stony mass), enclosing a sheltered lagoon. The circling reef rises above sea level in some places and that forms the land area of the atoll. Since coral can't live in cold water, coral atolls are found only in the tropics. Lagoons enclosed by these reefs vary in depth. Some are shallow, others have deep parts where large ships can anchor. But the greatest depths found in any lagoons seldom exceed 200 feet. In some lagoons, outcroppings of sharp coral rise to near the surface and are a hazard to vessels. Some lagoons have deep water passages through the reef; others are entirely enclosed. Often a passage through the reef can be opened with explosives. The size of lagoons can be anywhere from one mile to about 30 miles across.

Islets, which may be as much as 10 miles long, but which seldom exceed 400 yards in width, form along the flat tops of the circular coral reefs and on top of the scattered reefs inside the lagoon. These islets seldom rise more than 10 or 15 feet above sea level. Coconut palms, pandanus palms and similar vegetation grow thickly along the islets and some trees may rise to 100 feet. There is not much soil on these islets and what little there is has a high salt content, which prevents the growth of many tropical and temperate plants. In particular, grass will not grow, which means there is no vegetation on which cattle or horses could graze. But other salt-hardy vegetation does grow abundantly on many atolls, including the coconut palm which is very important in the economic life of small islands. Rocks on atolls are all limestone consisting of calcium carbonate produced by coral and shellfish.

There are no native animals, but introduced pigs and rodents are found on some atolls. Sea birds and their eggs are usually plentiful, and abundant fish and seafood are easy to catch in the lagoon. There is no running water. On populated atolls, rainwater is saved for drinking, and solar distillation of seawater should be feasible under the tropical sun. The main hazard of life on an atoll is tropical storms whose winds can have a devastating effect. And the climate is unpleasant at sea level in equatorial latitudes — hot and humid with little wind for relief. But, despite these deficiencies, many atolls do support large indigenous populations living a subsistence lifestyle.

Mere survival on an atoll is usually not a problem. Abundant food and building materials are available for the taking. It never gets so cold that artificial heat or much clothing is needed. Aside from storms, the main problems are boredom and lack of a way to earn much money to buy desired goods from outside.

Kahn, on page 156 of his book (see references) says this about life on a typical atoll: "A family of 6 on Kapingamarangi (an atoll in the Carolines) can get by nicely with $100 cash a year, and this at the going price for copra (dried coconut), ...any able-bodied man can make without overexerting himself. Kapingamarangans can obtain for free all the building materials, fish, fruits, and vegetables they want. With whatever cash they have, they mainly buy cloth, rice, sugar, corned beef, curry, pepper, and cigarettes."

Living On or Under Water

While it may seem like a strange idea, putting the watery area of lagoons to use as living space also bears consideration. Even those atolls whose land area is only a few dozen acres or less in extent usually have relatively large lagoons, often in the range of 30 to 50 square miles. So if a way could be devised to live on or under the water in the lagoon, and not just on the tiny land area of the reef, an atoll resident would have all the living space he could use. Even the deepest lagoons are only about 200 feet deep in their deepest parts, and "aquanauts" have already lived in underwater habitats at greater depths than that, which means that all lagoons are inhabitable using technology which currently exists.

The greatest threat to life one would face on a small atoll is from tropical storms with high winds and from "tidal waves" (tsunamis) whose winds and high seas have been known to strip an atoll bare of all vegetation. But the solution to this problem lies near at hand. Only a few feet below this maelstrom, beneath the water of the lagoon, relative calm prevails. So it seems that an appropriate strategy would be to build a storm shelter underwater in the lagoon. Then, when a tropical storm blows in, islanders could retire to the shelter and wait in safety for the storm to pass.

Or alternatively, the roles could be reversed and the underwater habitat could be used as the main residence, and the above sea level area of the atoll could be used for growing food, recreation, etc. The land area could be left in a nearly natural state, which would offer this additional advantage to anyone who wanted seclusion: it would be almost impossible to tell that someone was actually living there on the atoll.

But if living underwater is less than appealing, there are several ways to live on the surface of the water in a lagoon. One could simply live on a small boat anchored in the lagoon. Or one could live on a specially designed structure which, unlike a boat, would remain stationary, but which could be made larger and more stable than a boat. Such a structure would provide living space above and perhaps also below the water. Several designs for such aquatic habitats have been proposed.

Another more ambitious possibility is to use a large vessel to create an artificial island in the lagoon of an atoll. For example, one could bring an obsolete tanker with a large, flat deck into a suitable lagoon and partially sink it so that it rests on the shallow bottom with its deck well above the water. Soil could be spread on the deck and salt-hardy vegetation planted to create an instant island. For all of these schemes, the reef would serve as a natural breakwall providing protection against ocean waves.

For a typical example of what is possible, consider Taka, an uninhabited atoll in the Marshall Islands. Taka has a reef of 30 miles in circumference, but it has only five small islets with a total land area of 141 acres. The land area is too small for Taka to be settled by people intent on living in a conventional atoll lifestyle. But Taka's average-sized lagoon has an area of about 71 square miles, which is over 45,000 acres! Here is room enough for any size settlement, provided the settlers are willing and able to live on or under the water.

In the following pages we will visit deserted islands all over the world, from the tropics to the polar regions. Our purpose is mainly to describe the conditions that actually exist in these places. For further discussion about what could be done in such places along the lines of what has already been suggested, consult *The Last Frontiers On Earth* (see references) which describes in much more detail ways to live in polar regions and on and under water.

Part I
Pacific Ocean, Warm Islands

1. Hawaiian Islands

The US state of Hawaii consists of eight large islands and a string of small, uninhabited islands, called the Leeward Islands, which lie at wide distances off the northwest of the main group. Politically, the Leewards are part of Honolulu and in area they total less than three square miles. Most of the Leeward chain is part of the "Hawaiian Islands National Wildlife Refuge," which is a sanctuary for millions of seabirds, seals, turtles and land birds. There are almost no large trees in the Refuge. No one is allowed to enter the Refuge without permission from the US Fish and Wildlife Service. No one lives in the Refuge except for about 20 members of the US Coast Guard who are stationed on Tern Island in French Frigate Shoals.

All except one of the larger islands are populated. The populated Hawaiian islands are: Hawaii, Maui, Oahu, Kauai, Molokai, Lanai and Niihau. These Hawaiian islands are uninhabited:

a. *Kahoolawe.* Kahoolawe is the one large island with no population. It lies about 6 miles southwest of Maui at 21° north, 157° west, and it is 9 miles long, 6 miles wide, and 1477 feet high at its highest point. It has poor soil, though some cattle are grazed there. The US armed forces uses it as a target island. The island is leased to a private citizen.

b. *Kaula.* Kaula is a small, bare rocky islet 550 feet high, about 20 miles southwest of Niihau at 22° north, 161° west. It's a solid, irregularly shaped, completely barren rock. There is a light station there for the US Lighthouse Service.

c. *Nihoa.* Nihoa, formerly called Bird Island, is located 150 miles northwest of Kauai at 23° north, 162° west. It has Polynesian statues and 12 fertile acres that were once intensely cultivated. Nihoa's total area is 155 acres and its highest point is 910 feet.

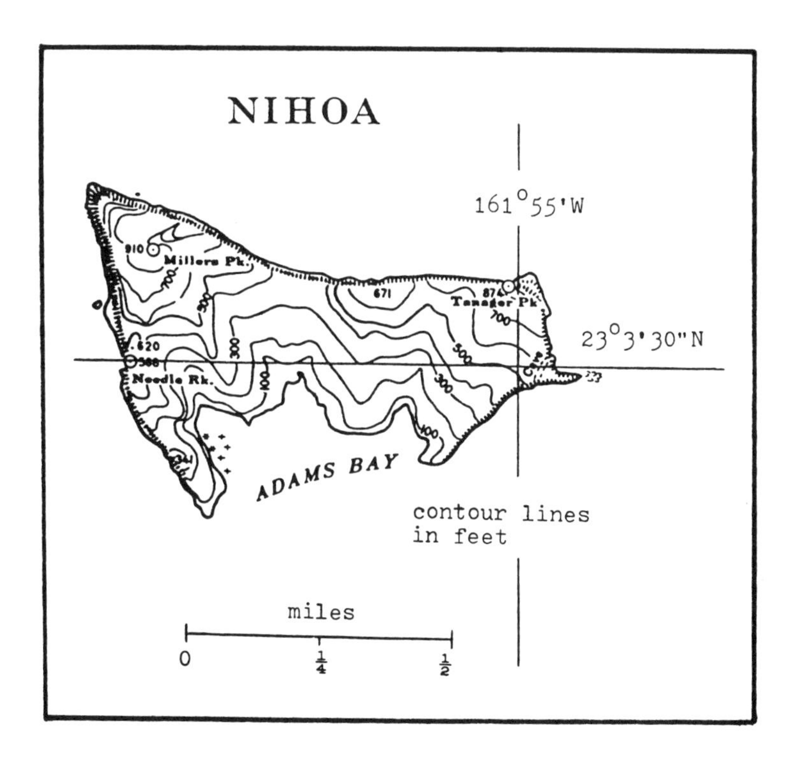

Nihoa

d. *Necker.* Necker is about 400 miles northwest of Honolulu at 24° north, 165° west and it is 277 feet high at its highest point.

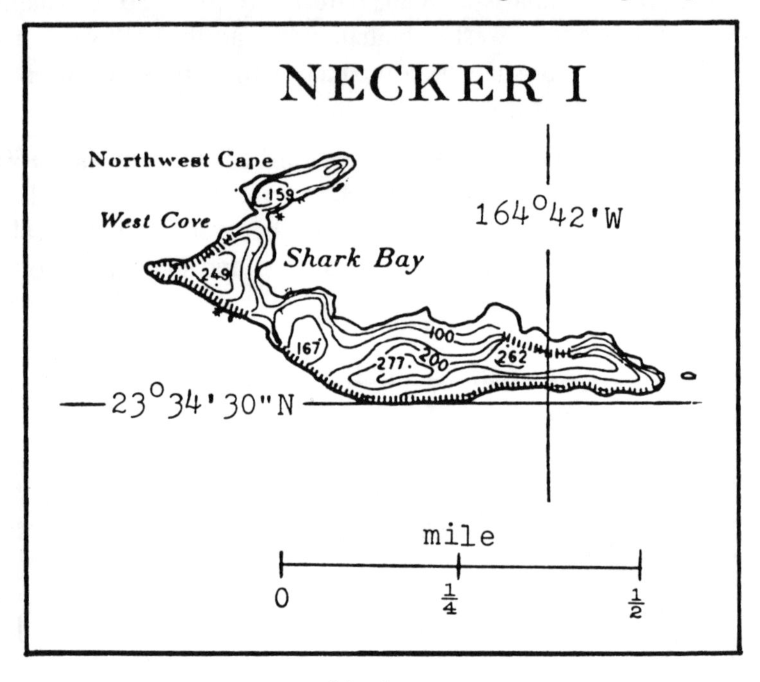

Necker

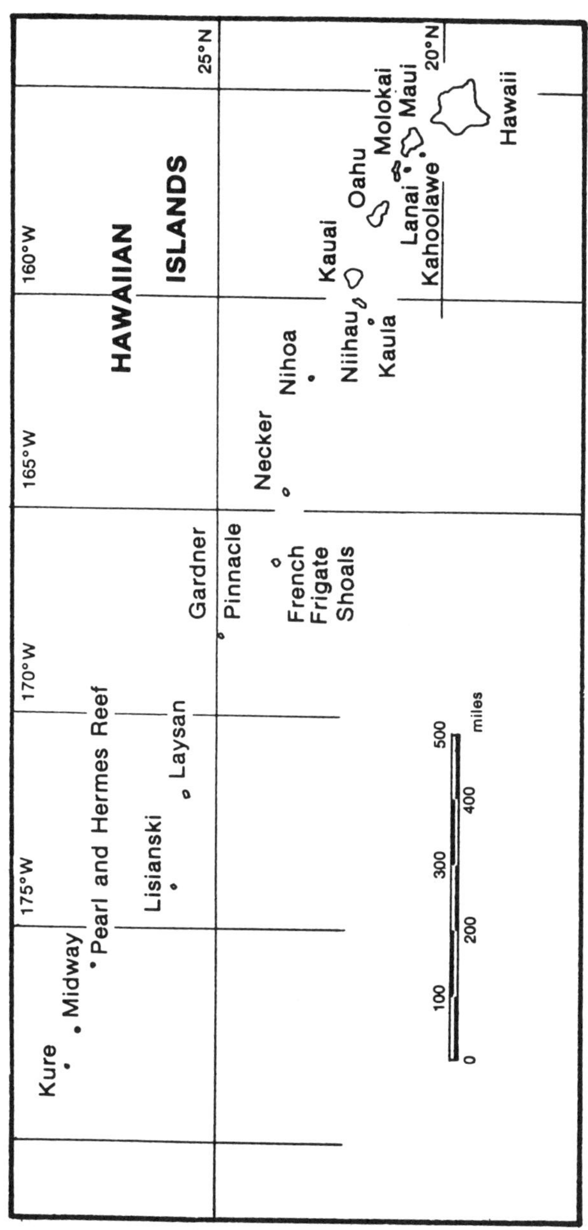

Hawaiian Islands

e. ***French Frigate Shoals.*** French Frigate Shoals is a crescent-shaped atoll consisting of 13 sand islets, including Whale-Skate, Tern, and East Islands, and one rock islet, known as La Perouse Pinnacle which rises to 122 feet. The atoll is located about 480 miles northwest of Honolulu at 24° north, 166° west. French Frigate Shoals was annexed by the Republic of Hawaii in 1895. Formerly, there was a US Coast Guard station on East Island, but that was abandoned in 1952. Presently, there is an active Coast Guard station on Tern Island on the site of a Navy airfield that was used in World War II. The Coast Guard base has a Loran station whose radio signals help ships and planes to compute their position. The landing strip is still used to bring in supplies.

La Perouse Pinnacle bearing Northeasterly.

La Perouse Pinnacle

f. *Gardner Pinnacles.* Gardner Pinnacles is about 590 miles northwest of Honolulu at 25° north, 168° west and is 190 feet high.

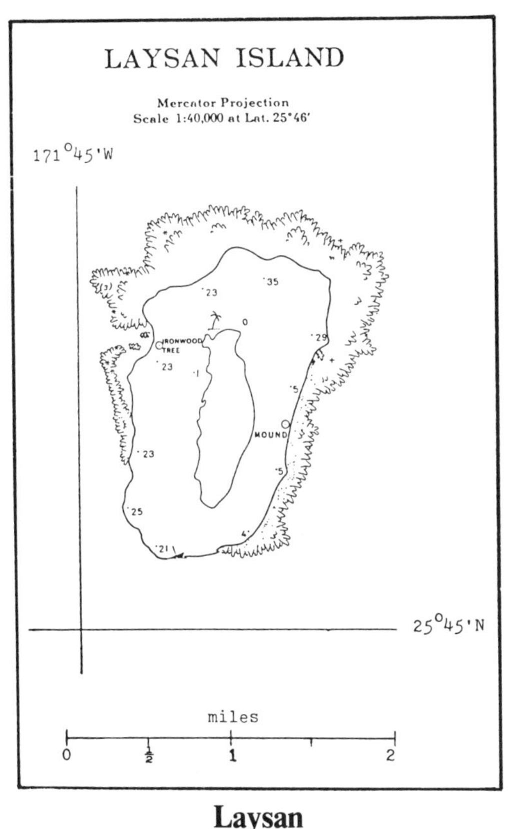

Laysan

g. *Laysan.* Laysan is located 790 miles northwest of Honolulu at 26° north, 172° west and is a sand island known for its numerous birds. It rises only 35 feet above sea level and is about 2 miles by one mile in size. In the past, Laysan has been mined for guano (bird droppings used as fertilizer), and birds' eggs and feathers (for hats) were gathered there at such a rate that the bird population was decimated. Early in the 20th century, rabbits were brought to the island. They destroyed the birds' habitat and 3 species became extinct as a result. In 1923, an expedition to Laysan exterminated the rabbits and the vegetation and bird life have since recovered.

One species, the Laysan Teal, a kind of duck, escaped extinction by the narrowest possible margin. The species had declined to the point where only one individual, a female, was left. She laid a clutch of eggs which were carefully nurtured and they survived and multiplied. Naturalists have brought the species back from the brink to the extent that the world population of Laysan Teal is now about 800.

In 1969, a Japanese fishing trawler ran aground on Laysan and the wreckage is still there.

h. *Lisianski.* Lisianski is a level, sandy island about 900 miles northwest of Honolulu at 26° north, 174° west. It is only 20 feet high.

i. *Pearl and Hermes Reef.* Pearl and Hermes Reef is an oblong atoll consisting of 7 sand islets, situated 1,050 miles northwest of Honolulu at 28° north, 176° west. It was discovered in 1822 and annexed in 1857 by the Kingdom of Hawaii. It was formerly worked for guano.

j. *Kure.* Kure, formerly called Ocean Island, is a circular atoll, about 20 feet high, which is the most northwesterly of the Hawaiian Islands. It can be found at 28° north, 178° west, about 1,300 miles from Honolulu, and about 50 miles northwest of Midway (which is a US possession, but not part of the state of Hawaii). Kure has been worked for guano. It was placed under the US Navy in 1936, but is now part of the state of Hawaii. Kure has a Loran (radio) station and a Coast Guard base whose commanding officer is in charge of the island.

2. Marcus

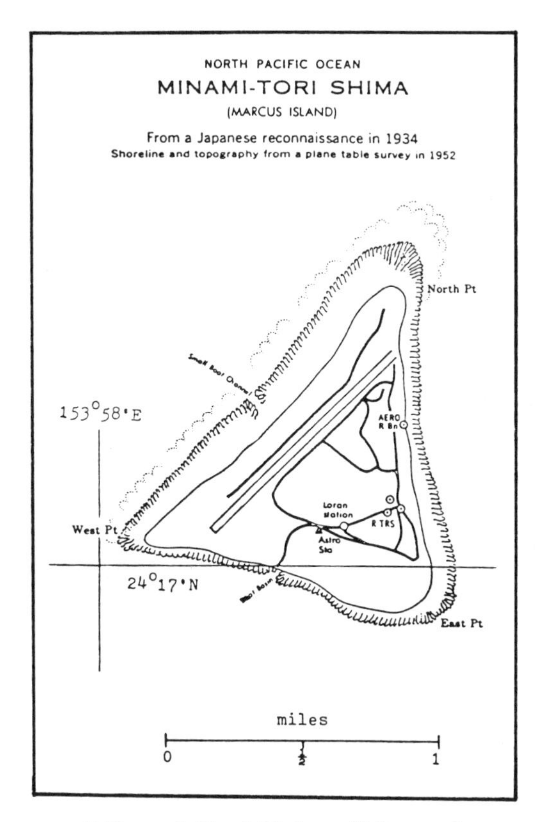

Minami-Tori Shima (Marcus)

Marcus, also known as Minami Tori Shima, is an isolated Japanese island located 850 miles east of the Bonin Islands at 24° north, 154° east. It is a triangular island of volcanic origin, about 2 miles long, with an area of about one square mile. The island rises to 204 feet and has some phosphate deposits. Prior to 1941, Japan built a powerful naval and air station on Marcus. The island was administered by the US after World War II, but was returned to Japan in 1968. It is now an albatross bird sanctuary.

3. Trust Territory of the Pacific Islands

The Trust Territory of the Pacific Islands was the part of Micronesia controlled by the US. It lies in the western Pacific and covers an ocean area about the size of the continental US. More than 2,000 islands totalling less than 700 square miles are scattered in several island chains: the Marshalls, Carolines, Marianas, and the Palauan Islands. These islands were taken from Japan by US armed forces during World War II. In 1947, the US organized these islands as a "strategic trust" under the United Nations trusteeship system. In 1980, after negotiations that extended over several years, the Trusteeship was terminated. As a result of these negotiations, the Trust was divided into 4 political entities. Three of these, the Marshall Islands, the Republic of Palau, and the Federated States of Micronesia (the Caroline Islands), entered into a "free association" with the US. The 4th unit, the Northern Marianas, sought and was granted closer ties to the US as a commonwealth.

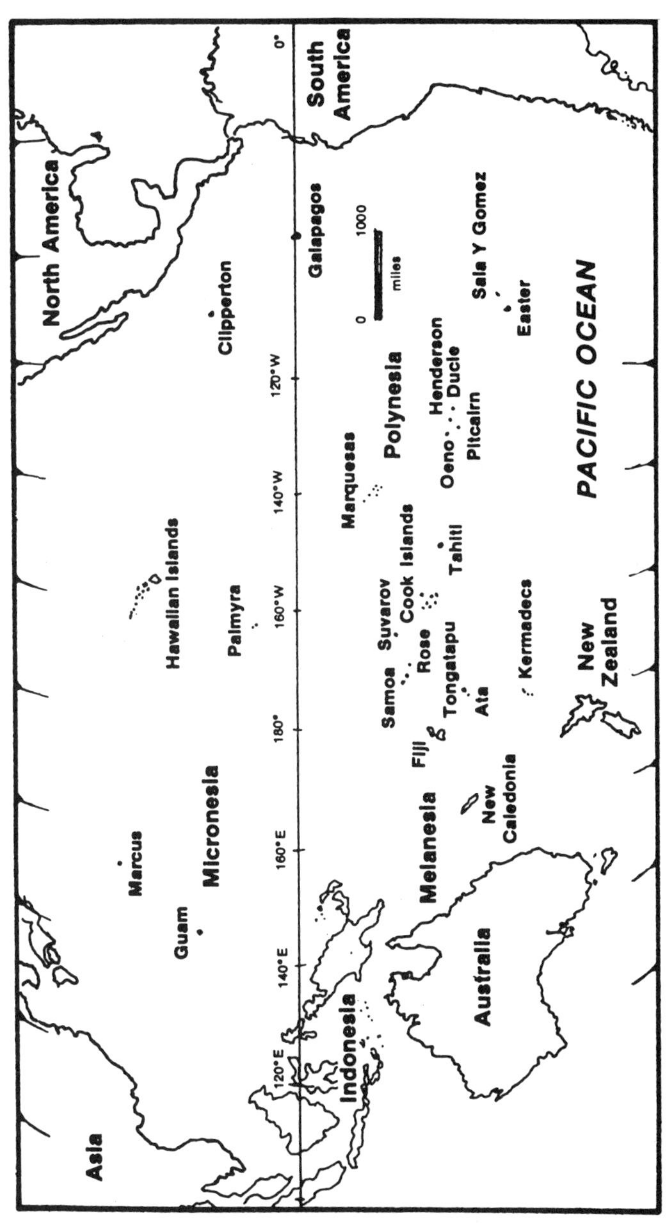

Pacific Ocean

4. Federated States of Micronesia

The Federated States of Micronesia came into being in May, 1979, with its capital on the island of Ponape. The new capital replaced the former Trust Territory capital of Saipan in the Northern Marianas. The Federated States of Micronesia consists of the states of Yap, Truk, Ponape, and Kosrae.

A. Yap

Yap consists of 16 island units, which include the Yap Islands proper, 5 single islands, and 10 atolls. Yap Island itself consists of 4 major and 6 minor high islands surrounded by reefs. About 65% of the state's population of 9,320 resides at Yap Islands. All the other islands of the state, except Fais, which is a raised atoll, are low lagoon type atolls. These 4 atolls of Yap state are usually uninhabited:

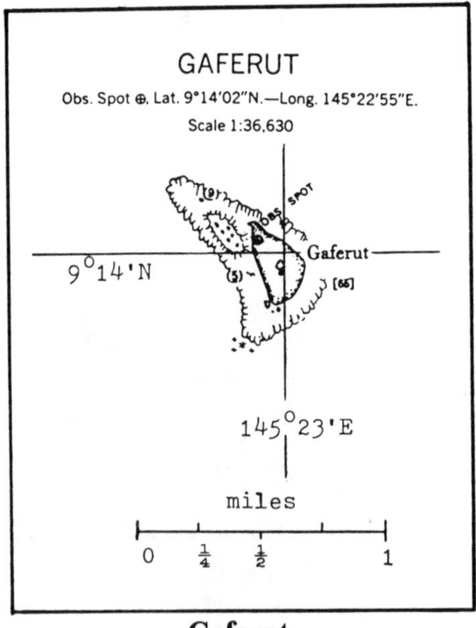

Gaferut

a. *Gaferut.* Gaferut is a coral atoll lying 60 miles northeast of Faraulep at 9° north, 145° east. It is ¾ mile long, ½ mile wide, and consists of one islet with an area of 28 acres. The island has mangrove trees, and rock phosphate deposits have been mined there.

b. *West Fayu.* West Fayu consists of one islet of 15 acres. It lies 46 miles northwest of Satawal, and is about 4 miles long by 1½ miles wide. West Fayu is owned by people who live on Satawal and anyone planning to visit West Fayu should get permission from the owners. The lagoon is very rich in fish, and people from Satawal make the journey to West Fayu from time to time to fish there.

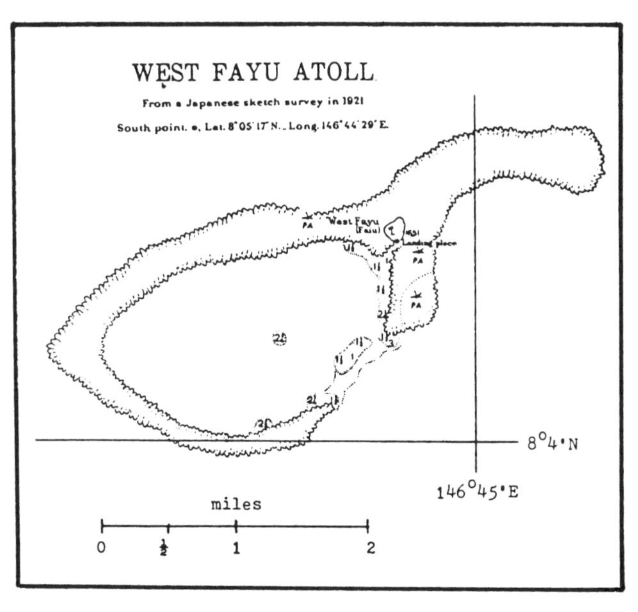

West Fayu

c. *Pikelot.* Pikelot is an atoll 2½ miles long and 1¾ miles wide, located 50 miles east of West Fayu. The atoll contains one islet with an area of 23 acres.

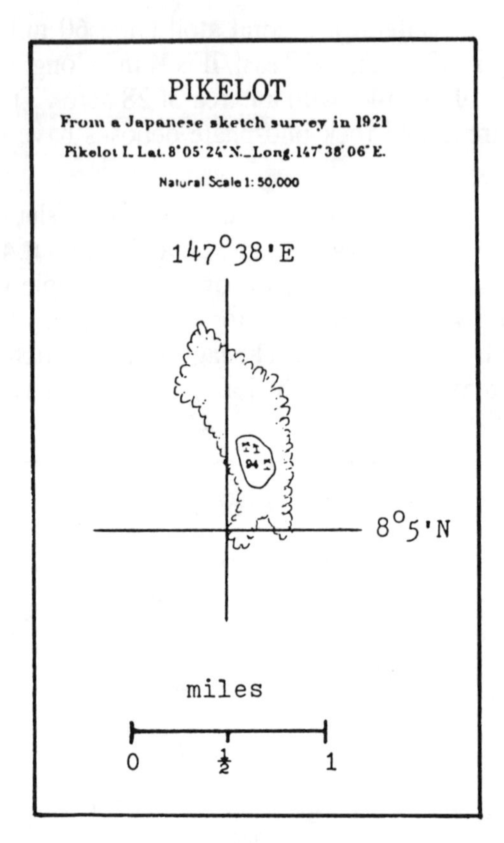

Pikelot

d. *Olimarao.* Olimarao lies 21 miles northwest of Elato at 8° north, 146° east. It consists of 2 wooded islets with a total area of 54 acres, and the dimensions of the whole atoll are 2 and one third miles long by 1½ miles wide.

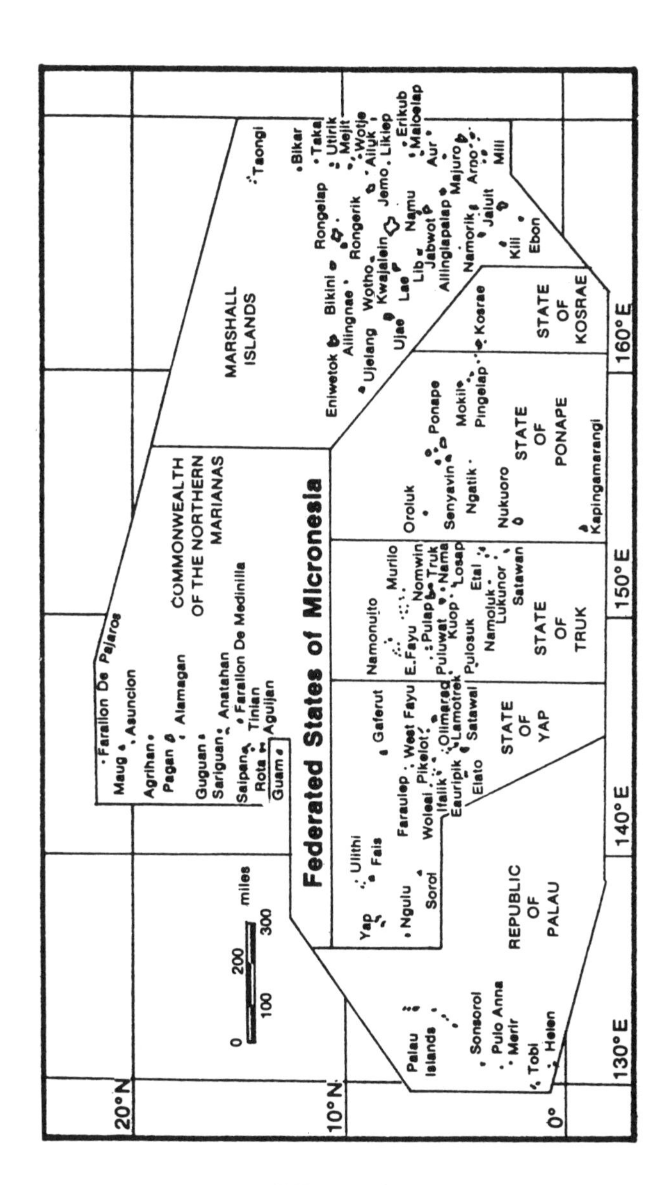

20°N
10°N
0°
130°E
140°E
150°E
160°E
0 100 200 300 miles
MARSHALL ISLANDS
COMMONWEALTH OF THE NORTHERN MARIANAS
Federated States of Micronesia
STATE OF YAP
STATE OF TRUK
STATE OF PONAPE
STATE OF KOSRAE
REPUBLIC OF PALAU
Farallon De Pajaros
Maug
Asuncion
Agrihan
Pagan
Alamagan
Guguan
Sariguan
Anatahan
Farallon De Medinilla
Saipan
Tinian
Aguijan
Rota
Guam
Taongi
Bikar
Taka
Utirik
Mejit
Wotje
Ailuk
Likiep
Erikub
Maloelap
Rongelap
Bikini
Rongerik
Jemo
Aur
Majuro
Arno
Mili
Eniwetok
Ailingnae
Wotho
Kwajalein
Namu
Lib
Jabwot
Ailinglapalap
Namorik
Jaluit
Kili
Ebon
Ujelang
Ujae
Lae
Kosrae
Ponape
Mokil
Pingelap
Oroluk
Senyavin
Ngatik
Nukuoro
Kapingamarangi
Murilo
Nomwin
Namonuito
E. Fayu
Pulap
Truk
Puluwat
Nama
Kuop
Losap
Pulosuk
Etal
Namoluk
Lukunor
Satawan
Gaferut
West Fayu
Faraulep
Pikelot
Olimarao
Lamotrek
Satawal
Woleai
Ifalik
Eauripik
Elato
Ulithi
Fais
Yap
Ngulu
Sorol
Palau Islands
Sonsorol
Pulo Anna
Merir
Tobi
Helen

Micronesia

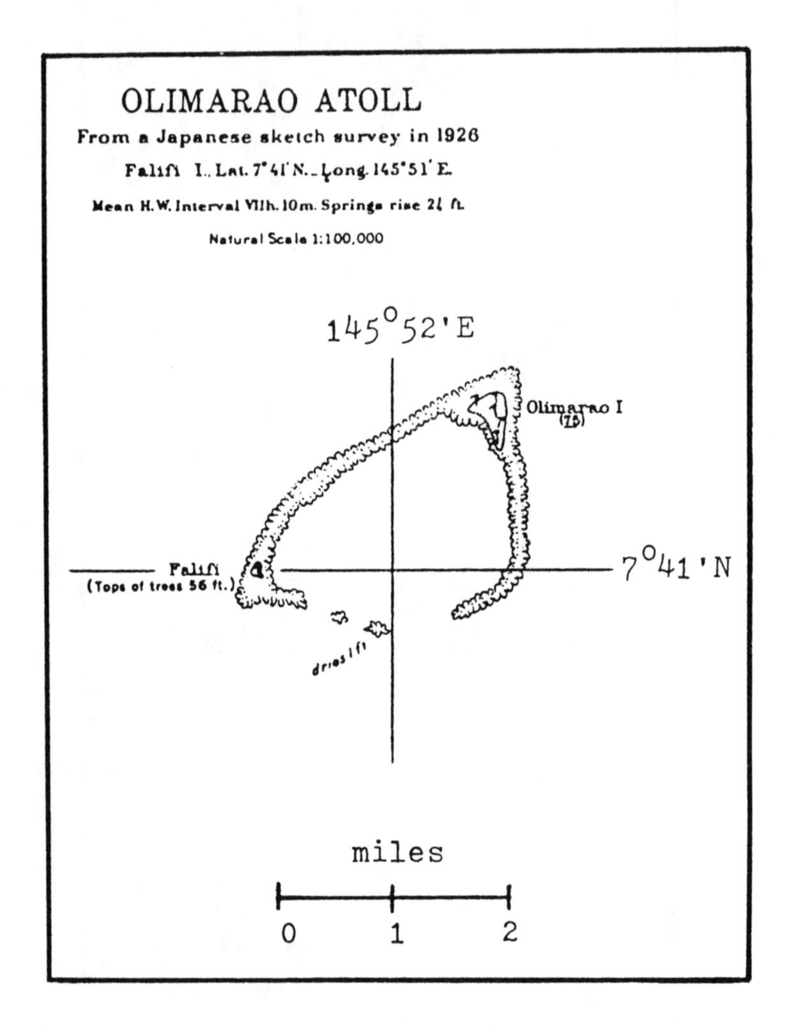
OLIMARAO ATOLL
From a Japanese sketch survey in 1926
Falifi I., Lat. 7°41'N.–Long. 145°51'E.
Mean H.W. Interval VIIh. 10m. Springs rise 2½ ft.
Natural Scale 1:100,000
145°52'E
Olimarao I
(75)
Falifi
(Tops of trees 56 ft.)
7°41'N
miles
0
1
2

Olimarao

B. Truk

Truk consists of 15 major island units. Except for Truk itself, all are low coral formations. Truk consists of 11 high volcanic islands enclosed by a coral ring broken into 87 islets. Two-thirds of the population of the state live on the 6 major high islands of Truk. The population of the state is about 40,000. Apparently none of the 15 major island units of Truk are uninhabited.

C. Ponape

Ponape consists of 9 island units, 8 low coral atolls and the high island of Ponape itself. Most of the state's population of 23,140 live on Ponape. None of the atolls have a population higher than 1,000. The one uninhabited island in the state of Ponape is:

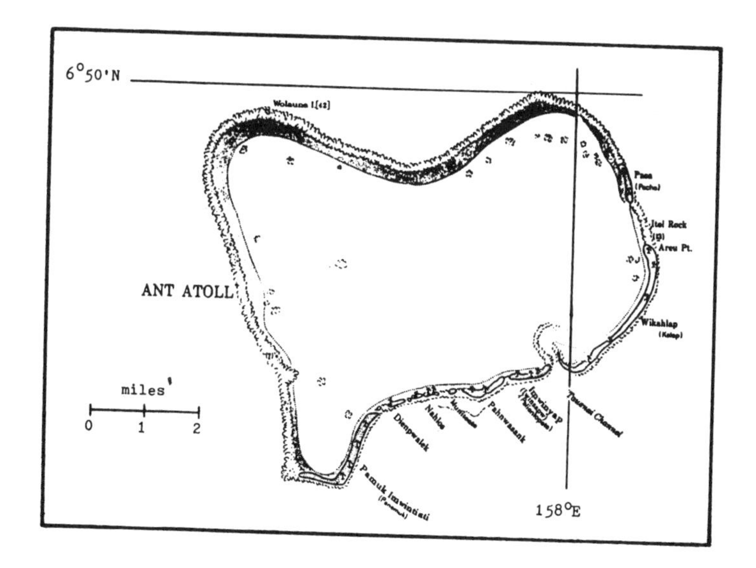

Ant

a. *Ant.* Ant, which lies 12 miles west of Ponape, consists of 2 large and 12 small islets with an area of 460 acres.

D. Kosrae

Kosrae state consists of a single high island of the same name which was formerly known as Kusaie Island. Kusaie used to be part of Ponape District of the Trust Terriroty but was made into a separate state when the Trust Territory was reorganized. The state of Kosrae has a population of about 5,000.

5. The Marshall Islands

The Marshall Islands are a group of 33 major island units, all low coral atolls, which were formerly part of the Trust Territory of the Pacific Islands. In July, 1978, Marshall Islanders voted against the proposed Micronesian Federation Constitution and decided to establish a separate autonomous government when the US trusteeship ends. A constitution was drawn up which was approved by the Marshall Islanders in a referendum held in March, 1979. In May, 1979, the new government established by this constitution began functioning. The total population of the Marshall Islands group is about 31,000. Only 3 atolls have more than 1,000 people: Arno has 1,095; Majuro has 5,602; and Kwajelein has about 8,000 Marshallese and 3,000 Americans. Kwajelein is the headquarters of the group. Its lagoon has an area of 1,100 square miles, which is the largest of any atoll in the world. Important US missle testing facilities are on Kwajelein.

The Marshalls are divided into two chains: the Ratak (sunrise) Chain to the east, and the Ralik (sunset) Chain further west. These 7 atolls in the Marshalls are uninhabited:

A. Ratak Chain

a. *Talongi.* Talongi (sometimes Taongi) is the northernmost of the Ratak Chain, 400 miles northeast of Kwajelein at 25° north, 169° east. It consists of 11 islets, is 15 feet high, and has an area of 1¼ square miles.

b. *Bikar.* Bikar is 250 miles northeast of Kwajelein at 12° north, 170° east. It is 5 miles long and contains 6 islets, with an area of 122 acres. Bikar also has guano deposits.

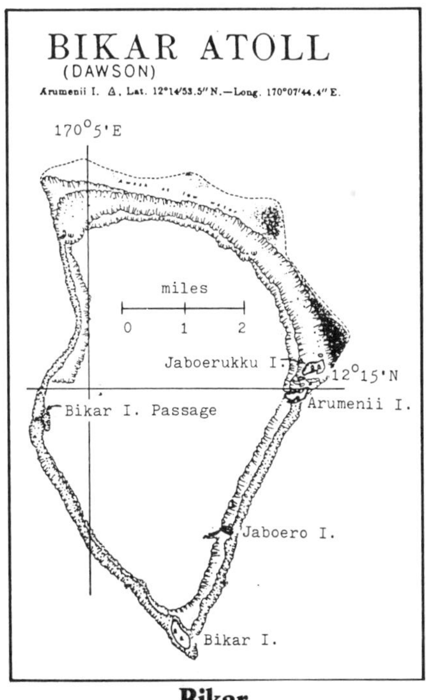

Bikar

c. *Taka.* Taka lies 190 miles northeast of Kwajelein at 11° north, 170° east. It is about 30 miles in circumference, consists of 5 islets, and has a land area of 141 acres.

d. *Jemo.* Jemo is an atoll 5 miles long with only one tiny islet with a land area of 38 acres. It is located 20 miles northeast of Likiep, at 10° north, 170° east.

e. *Erikub.* Erikub is located 175 miles east of Kwajelein at 9° north, 170° east. It is about 20 miles long and consists of 14 islets with an area of 378 acres.

B. Ralik Chain

a. *Ailingnae.* Ailingnae is a 15 mile long atoll consisting of 25 islets whose area totals about 690 acres. It is located 150 miles northwest of Kwajelein at 11° north, 166° east.

b. *Rongerik.* Rongerik lies 150 miles north of Kwajelein at 11° north, 167° east. It is about 30 miles in circumference, consists of 17 islets, and has an area of 416 acres. The residents of Bikini atoll were brought here by the US Government in 1946 in order to clear Bikini for atom bomb testing, but in 1947 they were moved again to Ujelang atoll.

6. Northern Mariana Islands

The Northern Mariana Islands were formerly part of the Trust Territory of the Pacific Islands. The group consists of 21 high, volcanic islands with a total land area of 182 square miles which extend in a line for about 400 miles. Most of the population of about 17,000 live in the southern section, the majority of them on the larger island of Saipan (14,600), Tinian (900), and Rota (1,275). The capital of the Northern Marianas is on Saipan.

Following the dissolution of the Trust Territory, the Northern Marianas became a Commonwealth of the US (a political status similar to Puerto Rico). Ceremonies on January 9, 1978, culminated over 20 years of effort by the people of the Northern

Mariana Islands to affiliate with the US. Regularly, from the start of the Trusteeship Agreement in 1947, the people of the Marianas had petitioned for closer ties with the US. When it appeared that the political status negotiations being conducted between the US and the Joint Committee on the Future Status of the Congress of Micronesia would not result in Commonwealth status for all of the Trust Territory, the Marianas delegation in 1972 requested separate political status negotiations with the US. This resulted in the signing of a Covenant between the US and the representatives of the Northern Marianas in February, 1975, which provided for the establishment of a Commonwealth of the Northern Mariana Islands. The signing was followed by a plebiscite held in the Northern Marianas on June 17, 1975, which approved the Covenant with 79% of the vote. Then the Covenant was submitted to the US Congress and was enacted on March 24, 1976.

Establishment of a Commonwealth means that the Marianas will have a closer relationship to the US than other parts of the Trust Territory. For one thing, Mariana Islanders will be US citizens and will be allowed unrestricted entry into the US, like Puerto Ricans. Other US citizens will no longer need entry permits for travel to the Northern Mariana Islands, However, only persons of Northern Mariana descent will be permitted to own land in the islands for 25 years after termination of the Trusteeship Agreement. But land may be leased by US citizens and by nationals of other countries for commercial purposes.

At the southern end of the Marianas lies Guam, which is geographically part of the island chain, but politically distinct. Guam is a separate possession of the US, and was never part of the Trust Territory. Guam is the largest and most populous island of the Mariana Chain of islands with a population of 106,000.

The 9 islands of the northern section of the Northern Marianas rise precipitously as mountain peaks of rocky, volcanic materials. Some of these peaks are active volcanoes, which have erupted in the 20th Century on Farallon de Pajaros, Asuncion, Pagan and Guguan. In fact, a volcano on Pagan erupted as recently as May 1981, and all 53 residents of the islands were evacuated by a passing Japanese freighter. This vulcanism, the rugged terrain, lack of soil and insufficient rain make these islands dry, barren and generally unsuitable for habitation. As a result, only 5 of the islands of the northern section are permanently inhabited, and even they have less than 250 residents altogether. The 4 uninhabited northern islands are:

a. *Farallon de Pajaros.* Farallon de Pajaros is the northernmost of the Marianas at 21° north, 145° east. It is about 512 acres in area, 1,096 feet high, and an active volcano.

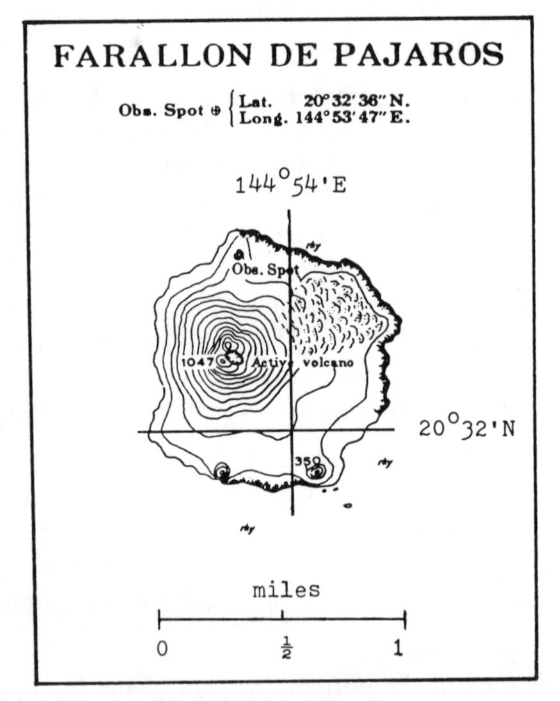

Farallon De Pajaros

b. ***Maug Islands.*** Maug Islands are a group of 3 steep peaks at 20° north, 145° east, connected by a common base beneath the water, surrounding a lagoon which is a sunken crater. The highest peak is 745 feet, and the area is about 512 acres. Uninhabited now, it was formerly a Japanese weather station.

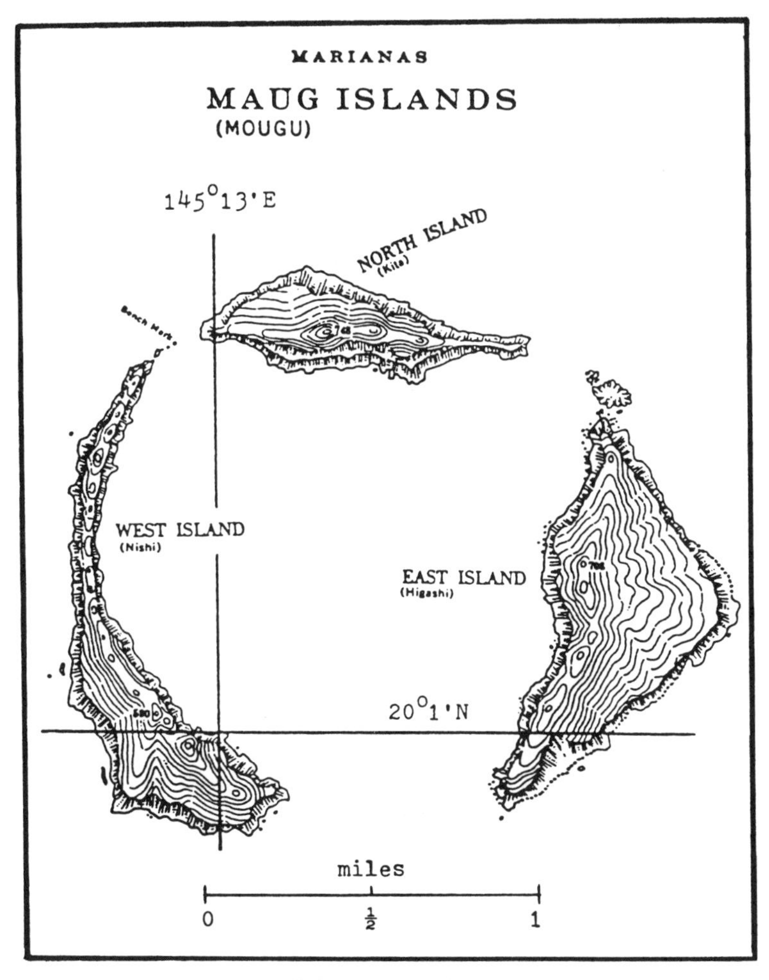

Maug Islands

c. *Asuncion.* Asuncion is a high island of volcanic origin, about 300 miles north of Saipan at 19° north, 145° east. It is about 2 miles in diameter, about 3 square miles in area, has a cone-like shape, and rises to a height of 2,923 feet.

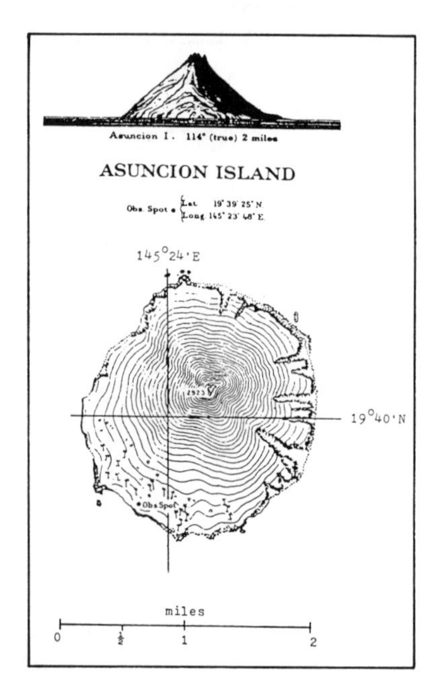

Asuncion

d. *Guguan.* Guguan, at 17° north, 146° east, is another active volcano, with twin peaks, the higher of which rises to 988 feet. It has an area of 1.6 square miles.

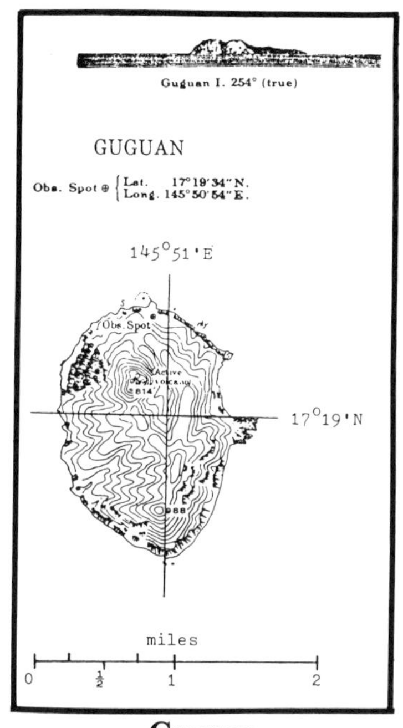

Guguan

e. *Aguijan.* In addition to the 4 uninhabited northern islands, there is one more uninhabited Marianas Island, Aguijan, the southernmost island of the Commonwealth. Also known as Goat Island, it lies 5 miles southwest of Tinian, at 15° north, 145° east. Oval-shaped Aguijan, 3 miles by one mile in size, consists of 3 concentric limestone terraces, surrounded by 100 foot high vertical cliffs on all sides. There is one difficult landing place, a steep ravine on the west side. Aguijan is not volcanic and it has sufficient moisture to support patches of forest and grassy meadows. The island's animal life includes feral goats, monitor lizards, rats and coconut crabs. It was used by the Japanese during World War II as a pineapple and sugarcane plantation. No one lives there now, but it is visited frequently by hunters from Saipan and Tinian who round up goats and gather crabs. These days, they usually come and go by air, landing on a 1,300 foot runway that runs along the highest central plateau of the island.

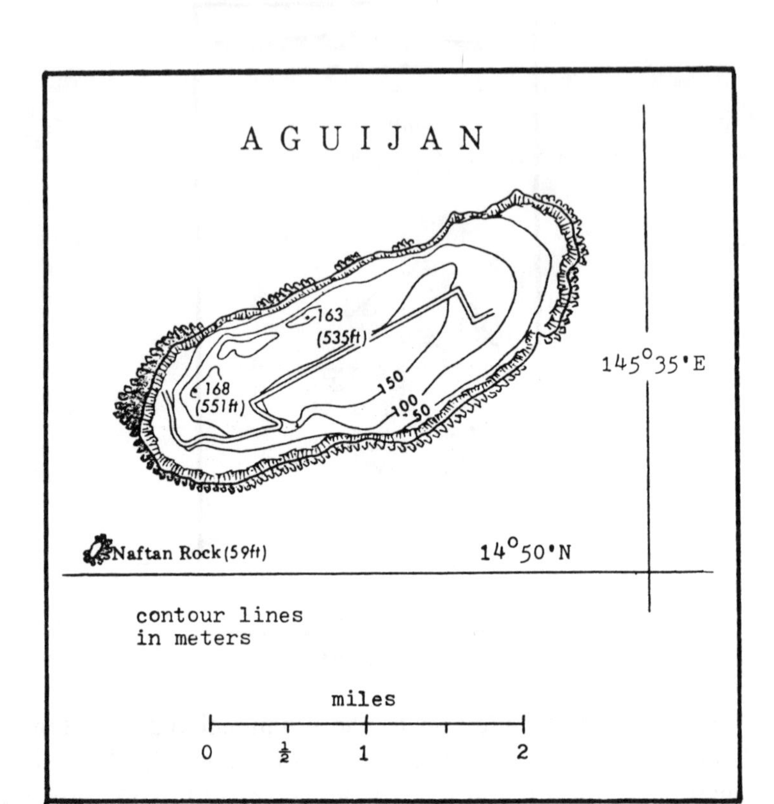

Aguijan

6. Republic of Palau

The Palauan Islands consist of one major island unit and 5 other small, isolated islands, which formerly comprised Palau District of the Trust Territory. In July, 1978, Palauans voted down the proposed Micronesian Federation and some months later they convened a Constitutional Convention to draw up a constitution for an autonomous Palauan government. Various political disputes delayed putting the new constitution into effect, but by January, 1981, the constitution had been approved by the voters and a government had been elected and the Republic of Palau became a functioning political entity.

The geographical heart of the Republic is the Palau cluster, which contains 4 high islands, one atoll and 338 tiny rock islets which are all part of one geographical formation. Of the total population of 14,800, less than 200 live outside the Palau cluster. The 5 other island units of the group are tiny, isolated islands lying in a line from the Palau cluster almost to New Guinea. All of these are single islands of raised limestone formations except Sonsorol, which has 2 components. Three of them are inhabited and have these populations: Sonsorol has 95 inhabitants, Tobi 75, and Pulo Anna has 13. These remaining 2 are uninhabited:

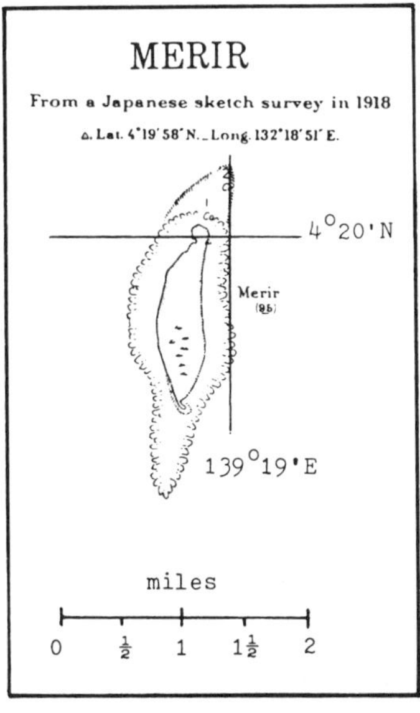

Merir

a. *Merir.* Merir is about 29 miles southeast of Pulo Anna at 6° north, 132° east. It is one and one-third miles long, ¼ mile wide, has an area of 223 acres, and rises to 50 feet.

b. *Helen.* Helen reef lies about 45 miles east of Tobi at 3° north, 132° east. It consists of one small islet of 487 acres on an oval reef that's about 12 miles by 30 miles. The reef completely surrounds a lagoon except for one channel that's open to the sea. At low tide, most of this reef is one to 3 feet above water, but it's entirely awash at high tide.

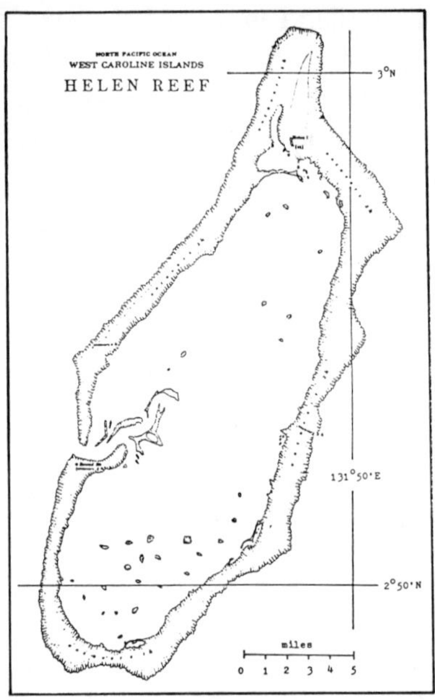

Helen Reef

8. Kiribati

On July 12, 1979, the new island nation of Kiribati was born. It is made up of 33 islands in the mid-Pacific lying in the Gilbert, Phoenix and Line Island chains, with its capital at Tarawa in the Gilberts. Although Kiribati is spread over 2 million square miles of ocean, its land area totals only 264 square miles. The population of these islands is about 65,000.

Formerly, Kiribati had been the British Colony of Gilbert and Ellice Islands, a British possession since 1892. In 1937 the Phoenix Islands were added to the colony. The 3 populated Line Islands of Christmas, Fanning and Washington were added to Kiribati because they are adjacent and are inhabited mainly by imported Gilbertese laborers. In 1975, the Ellice Islands split off from the Colony and they achieved independence in 1978 under the name Tuvalu. A year later Kiribati became independent.

A Treaty of Friendship negotiated between Kiribati and the US is now awaiting ratification by the US Senate. According to this treaty, the US claim to several islands in the Phoenix and Line Islands will be given up in favor of Kiribati. Islands to be ceded include all the Phoenix Islands to which the US has laid claim including the largest, Canton, and in the Line Islands: Christmas, and the southern Line Islands of Flint, Malden, Caroline, Starbuck and Vostok.

9. Line Islands

Running southeast from Hawaii to Tahiti, from 6° north to 11° south, by 151° west to 162° west, is a line of small, low, semi-barren islands with no native inhabitants, and almost no value apart from deposits of guano and coconut plantations. In the late 19th and early 20th centuries some of these islands were worked for guano, and small planting companies established coconut plantations in some of the more accessible places. In the 1930's the Line Islands were considered for air bases for trans-Pacific flights, but today's longer range aircraft no longer need mid-Pacific bases.

From north to south the Line Islands consist of: the US posessions of Kingman Reef, Palmyra and Jarvis; the inhabited islands of Christmas, Fanning and Washington, which are part of Kiribati, formerly the British Colony of Gilbert and Ellice

Islands; then the uninhabited islands: Starbuck, Malden, Flint, Vostok and Caroline, which had been claimed by both the US and Britain, but which are now being ceded to Kiribati.

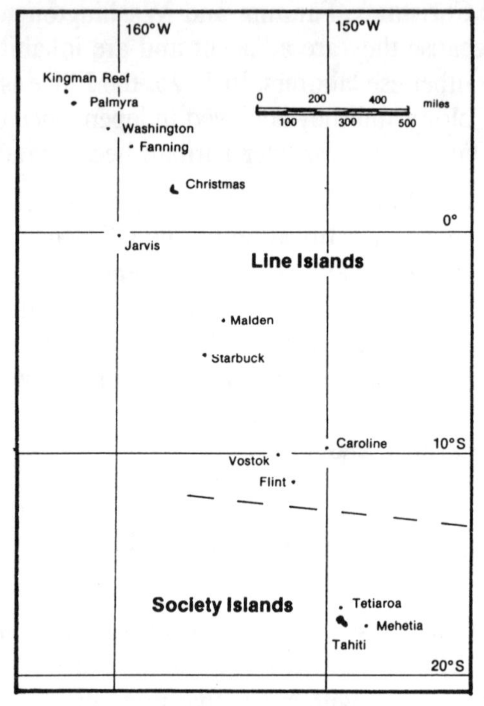

Line Islands and Tahiti

a. ***Kingman Reef.*** Kingman Reef is a triangular formation at 6° north, 162° west, about 920 miles south of Hawaii, and 35 miles northwest of Palmyra. It is a bare reef, 9 by 5 miles, running north and south the long way, with one corner of the triangle pointing north. Inside the reef lies a lagoon of considerable depth. The only land area is on the east side, where a small islet rises about 3 feet above sea level at high tide. Kingman was discovered in 1798 by the American Captain Fanning, but it was named after Captain Kingman, who visited the reef in 1853. It was annexed by the US in 1922 and made a US Naval reservation in 1934. In 1937, Pan Am began air mail service between Hawaii and New Zealand using flying boats, and they used Kingman Reef's sheltered lagoon as a stopover. A schooner was anchored there as a sort of hostel and supply ship. The route was abandoned after a seaplane was lost off Samoa in 1938. Kingman Reef is still under the control of the US Navy. There are many reefs similar to Kingman Reef throughout the Pacific and other oceans.

b. *Palmyra.* Palmyra is a low, wooded atoll, located at 6° north, 162° west, about 1,000 miles southwest of Honolulu. It consists of about 50 small islets totaling 250 acres, surrounding 3 lagoons that run down the center of the atoll. A platform of coral and hard sand runs around the islets and lagoons and, even at high tide, one can walk from one islet to another. For the most part these islets are only 5 or 6 feet above sea level with the highest elevation of 30 feet, but dense vegetation, including coconut trees and balsa-like trees, rises to a height of 100 feet. In some places this vegetation forms an almost impenetrable jungle of trees and vines. The whole atoll is 5 miles long by 1½ miles wide. Palmyra has no sheltered harbor and landing by sea is difficult owing to the large number of coral heads in the approaches. There is an abandoned airfield on the atoll.

Palmyra was discovered in 1802 by Captain Sawle of the American ship *Palmyra.* It was claimed by King Kamehameha IV for the Kingdom of Hawaii in 1862, then by Britain in 1889, and was included by the US among the Hawaiian Islands in 1898. Judge Cooper of Honolulu acquired title to the island in 1911, and used it for growing coconuts. Later he sold all except 2 islets to Leslie and Ellen Fullard-Leo. It is now owned by Ainsley Fullard-Leo of Kailua, Hawaii. Cooper died in 1929 and his 2 remaining islets (Home Islands) passed to his heirs.

An important US naval air transport base was developed on the atoll during World War II, at which time the island was a "prohibited defense area." Beginning in 1939, a coral airstrip was built from reef rock, which increased the land area of Palmyra to about 1,000 acres. The lagoons were dredged at that time to create a harbor and seaplane landing area. During the war, up to 6,000 military personnel were stationed on the atoll, and various facilities, such as fortifications, roads and causeways, were built on the island. At this time a Pan American Hotel operated on Palmyra, but the hotel and other buildings were destroyed by fire in February, 1948, at about the time the government was evacuating the island. There's not much left now of the multi-million dollar military investment on Palmyra, except for the dredged channel, concrete bunkers and the overgrown remains of the airstrip.

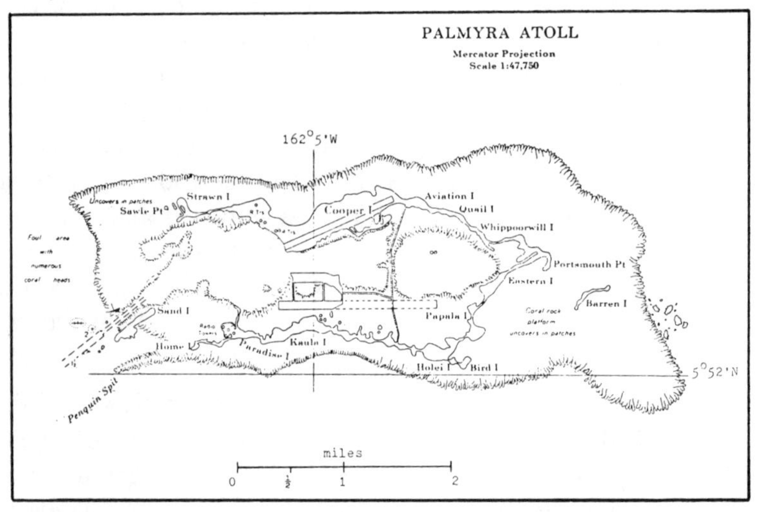

Palmyra

After the war, the U.S. Government tried to hold on to the atoll, but in 1947 the US Supreme Court ruled against the Government's claim to the island and confirmed its private ownership. In 1961, an Executive Order placed administration of Palmyra under the Secretary of the Interior, although it remains privately owned. The atoll is not part of the State of Hawaii. Kahn (see references) reports that the family that owns Palmyra incurred the wrath of a US Senator, so when Hawaii became a state, the Senator managed to get Palmyra excluded to prevent the owners from deriving any benefit from statehood. The atoll is presently uninhabited, except for an occasional visit by yachts, and it is for sale for $16 million.

The Soviet Union maintains a missile test range in the Pacific Ocean south of Hawaii near Palmyra, and they warn ships to stay away from the area while they are firing test shots. Soviet naval vessels patrol the area during tests to collect data on the missiles as they splash down.

In August, 1979, the US Government announced that it had selected Palmyra to be the site of the first radioactive waste dump in the Pacific. Palmyra edged out Midway and Wake Islands which were also under consideration for this dubious honor. According to the plan, 100 acres on Palmyra would be used as a dump for nuclear garbage from Japan, Korea, Taiwan and the Philippines, and radioactive material would begin arriving in 1986. The 100 acre dump site is considered capable of storing 30,000 tons of radioactive wastes. The wastes would be placed inside steel containers covered with concrete. These containers are expected to last only thirty years and the plan calls for moving the wastes to some other underdetermined dump site at the end of that time.

The private owners of Palmyra were outraged by this announcement and they pledged that they would not sell Palmyra to the US Government for any price, and they would not allow anyone to dump radioactive garbage on the atoll.

c. *Jarvis.* Jarvis is a small, bleak, bowl-shaped island 2 miles by one mile, with an area of 1.7 square miles, lying by itself at 0° latitude, 160° west. It has a narrow fringing reef around the shoreline, but no lagoon. There is a spot where a landing can be made, but the island lacks a good anchorage. It has no fresh water and gets little rainfall. Jarvis was worked for guano from 1857 to 1879 by the American Guano Company, annexed by Britain in 1889, then claimed and occupied by the US in 1935 with no objection from Britain. At various times since then it has been used as a weather station. During these occupations, a settlement named Millersville was built on the western side of the island, consisting of a number of wood and stone houses, with a radio shack and towers. Jarvis was occupied for a time during the International Geophysical Year (1957/58). It is now under the control of the US Fish and Wildlife Service as part of the National Wildlife Refuge System.

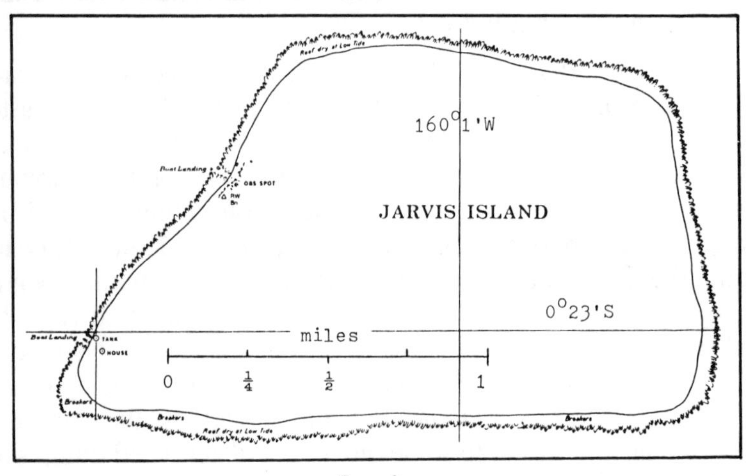

Jarvis

d. *Starbuck.* Starbuck is 336 miles south of the equator and 108 miles southwest of Malden at 6° south, 156° west. It is low, about 8 to 10 feet in height, 5½ miles long by 2 miles wide, with less than 1 square mile of land area. It is barren and treeless and has a very white beach in the shape of an angle. A reef encloses shallow lagoons. Arundel and Company began digging guano from the island about 1870, and during this period an opening for small boats was made in the reef on the northwest side. The island has been unworked and unoccupied since 1920. Attempts to plant coconuts failed. The British built a beacon on the western end as a warning to shipping.

e. *Malden.* Malden is 240 miles south of the equator at 4° south, 155° west, about 430 miles southeast of Christmas Island. It is a triangular, flat, coral island, on a reef measuring 5 miles by 4 miles, which encloses a large lagoon. It is a barren place where only stunted vegetation grows. But early European visitors saw stone faced platforms and graves which indicate that Polynesians lived there long ago. From about 1849, guano deposits were worked by Australians for 70 years. The island has been unused and unoccupied since 1927. More recently some notice was given to the island in connection with the British atomic and hydrogen bomb test on Christmas Island during the period 1956 to 1962.

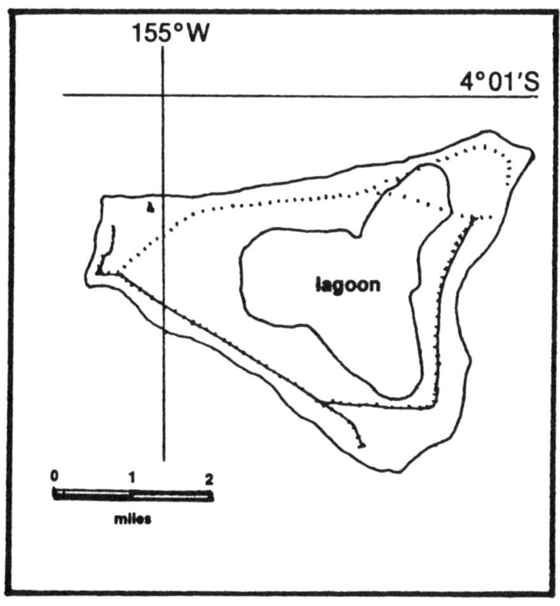

Malden

f. *Caroline.* Caroline lies about 500 miles south of Tahiti at 10° south, 150° west. It consists of a coral reef about 6 miles long by one mile wide enclosing a shallow lagoon. There are 22 islets strung along the reef, some of them rising up to 20 feet above sea level. They are heavily overgrown with coconut palms, pandanus, and similar coral atoll vegetation. There is a bay on the western side, but no good anchorage offshore, and only small boats can get thru the reef into the lagoon. Water can be obtained by digging.

The atoll was once inhabited by Polynesians, and there was a village of about 30 people on the southern islet in 1868. It has been worked for guano, but the deposits are now all worked out. The populated Kiribati island nearest to Caroline, and the administrative center for the Line Islands, is Christmas Island, almost 1000 miles north of Caroline. The capital of Kiribati on Tarawa lies nearly 2,700 miles northwest of Caroline. In his book *Micronesia Handbook,* author David Stanley says that the extreme isolation of the Line Islands "has created serious transport and administrative problems for the present government." So the government is weak and very far away. A government that has "serious administrative problems" won't have the ability to closely watch uninhabited specks of land at the farthest extreme of its national territory.

g. *Vostok.* Vostok lies about 125 miles to the west of Caroline at 10° south, 152° west. It is a low, triangular, coral lump, a small compact island about ½ mile long by ½ mile wide. Vostok is distinguished only by a dense central clump of buka trees, 80 feet high. There is a boat passage through the reef at the southwest corner, but no anchorage. Vostok was worked sporadically for guano and copra from 1873 to 1943. It is now unoccupied and unworked.

h. *Flint.* Flint can be found at 11° south, 151° west, about 125 miles southwest of Caroline. It is a flat atoll, 3 miles long by ½ mile wide. It rises to a maximum height of 25 feet, is well wooded and is entirely surrounded by a coral reef, through which a boat passage has been blasted, leading to a landing place on the west side. The island was leased to Arundel and Company who worked it for guano. The guano deposits are now all worked out. Then it was leased to Maxwell and Company of Tahiti, whose plantations there contained about 30,000 coconut palms. A white manager and some 30 natives worked the plantations and were visited occasionally by a schooner from Tahiti. In 1951, Flint was leased to M.P.A. Bambridge of Papeete, Tahiti, in connection with Caroline.

10. Phoenix Islands

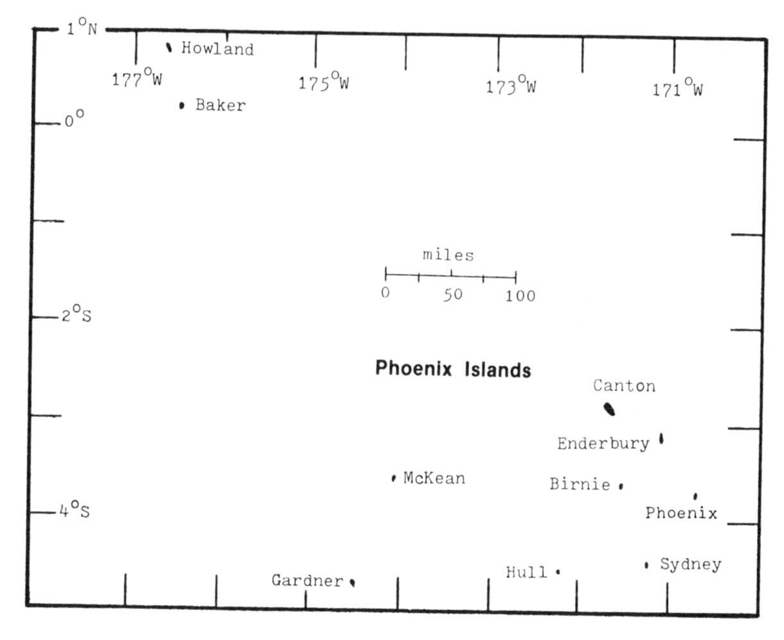

Phoenix Islands

The Phoenix Islands are a group of 8 small coral islands located at 3° south, 171° west, including Canton, Gardener and Hull, which are inhabited, and Enderbury, McKean, Phoenix and Sydney, which are not. Canton, the largest of these islands, along with Enderbury, was placed under an Anglo/American condominium in 1939. The other islands were annexed in 1937 to the British Colony of Gilbert and Ellice Islands, while the US continued to press its disputed claim. Canton Island was used as a trans-Pacific airline base in the 1930's, but now that trans-Pacific flights are made by longer range planes, they no longer stop at Canton. Many of these islands were formerly worked for guano, and some of them now produce copra. Ratification of the Treaty of Friendship between the US and Kiribati by the US Senate will complete the transfer of all of these islands to Kiribati.

Beginning in 1937, the British moved surplus population from the overcrowded southern Gilberts to the then uninhabited islands of Hull, Sydney and Gardner. The settlements on Hull and Gardner have increased to over 1,000 on each, but problems have developed on Sydney, and all of the settlers have been taken off and moved to the Solomon Islands.

a. *Enderbury.* Enderbury is a coral outcrop located at 4° south, 171° west, about 37 miles southwest of Canton. It is about 3½ miles long by 1½ miles wide, covered with scrub, with no satisfactory anchorage. Enderbury is mostly land, with the lagoon reduced to a shallow puddle. Both Enderbury and Canton are low and flat, for the most part bare, or covered only by low plants and grass. But each has a few clumps of tournefortia or tree heliotrope and a few stunted kou thickets.

Enderbury was discovered and reported by J.J. Coffin, a British captain in 1823, and named after a London merchant. The US Exploring Expedition under Commander Wilkes visited the island in 1840 and again in 1841. Various British and

American interests worked the large guano deposits between 1860 and 1890, but the island was not valued highly because it lacked an anchorage. There are no sea or land plane facilities. There was a small settlement and lighthouse on the southwest corner of the island. North of the settlement is a difficult landing place. Four American colonists were installed on the island in 1938, although British colors were still flying on the uninhabited island. The colonists were removed in 1942 after the outbreak of war with Japan. The island is now uninhabited.

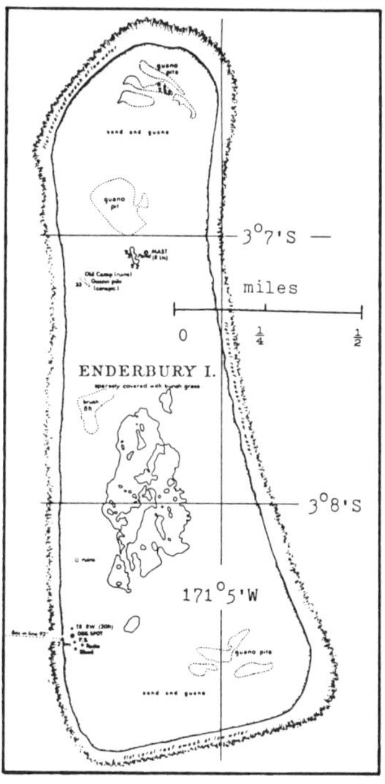

Enderbury

b. *Sydney.* Sydney is a roughly triangular island, with sides of 1¾ miles long, and a base about 2 miles across, located at 4° south, 171° west. Its shrunken lagoon has no outlet to the sea. Captain Emment found it in 1823, but it was neglected until

1882 when John T. Arundel and Company leased it. The British flag was raised on the island on June 26, 1889. After the turn of the century, Lever Brothers maintained coconut plantations there, and later Captain Allen cut copra for the Samoa Shipping and Trading Company. Polynesian ruins have been found on Sydney.

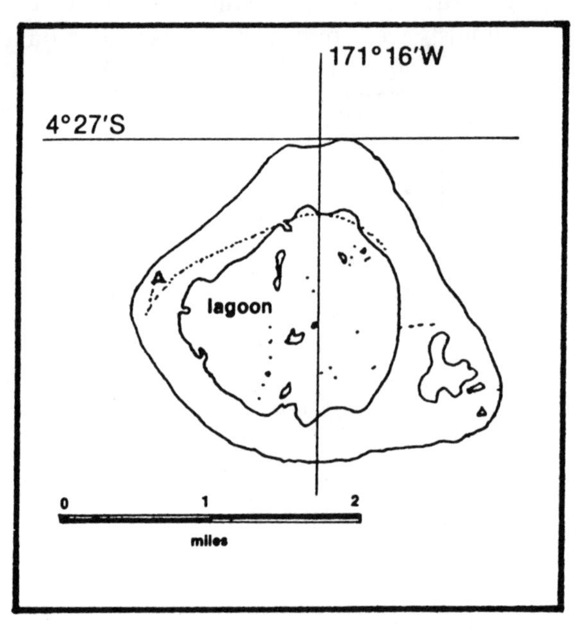

Sydney

Beginning in 1937, the British attempted to bring in settlers from the overcrowded southern Gilberts, but the settlement on Sydney was not successful. The trouble appears to have been due to the landlocked lagoon which, as it evaporates, is getting more and more salty, so that fish can not live in it, and it is apparently polluting the subsoil, thus harmfully affecting the growth of vegetation. All the settlers have been taken off and Sydney is now uninhabited.

c. *McKean.* McKean lies at 4° south, 174° west, about 175 miles southwest of Canton. It is oval in shape, ½ mile long by about 800 yards wide, for a total of 142 acres. The highest point of land rises to 17 feet on the north side. McKean was discovered in 1840 by Lieutenant Wilkes of the USS *Vincennes.* It received much attention from guano diggers, particularly the Phoenix Guano Company, until 1870. When the British Pacific Islands Company obtained a lease in the 1890's, there was no guano left worth shipping.

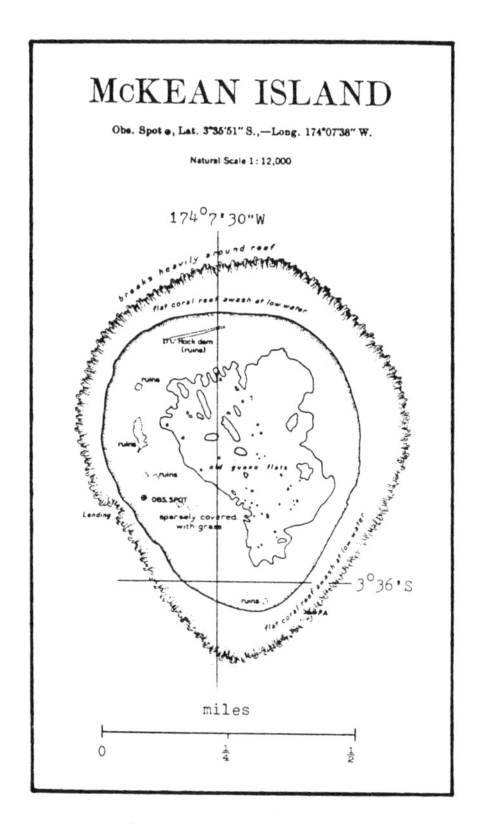

McKean

d. *Birnie.* Birnie lies 45 miles south of Canton at 4° south, 172° west, and is the smallest of the Phoenix Islands. It measures about ¾ mile long and runs to 600 yards at its widest part, for an area of 44 acres. Birnie is a low lying coral island with a shallow brackish lagoon and little else. There's a 20 foot high navigation beacon on the eastern side. Birnie was discovered and named by Captain Emment in 1823. The Pacific Islands Company held a lease on the island in 1899.

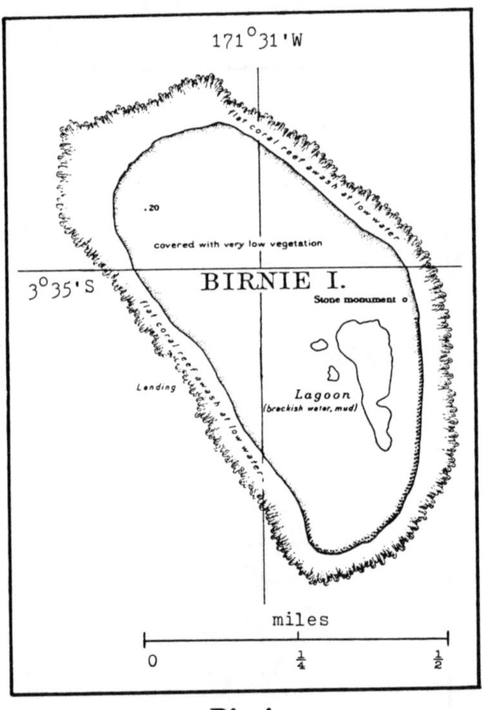

Birnie

d. *Phoenix.* Phoenix is a pear-shaped island less than ¾ mile long by ½ mile wide, with an area of 121 acres. It is not quite 20 feet high at its highest point. It lies 80 miles southeast of

Canton at 4° south, 171° west. It is the most fertile of the island group and is overrun with rabbits. It has fresh water pools in the center. There is no anchorage, but a landing may be made on the southwest side in moderate weather. It was discovered by the American ship *Phoenix* prior to 1828. Between 1859 and 1871, US interests, including the Phoenix Guano Company, worked the guano deposits. Great Britain annexed the island on June 28, 1889. Hundreds of thousands of sea birds nest on the island, which is now a bird sanctuary.

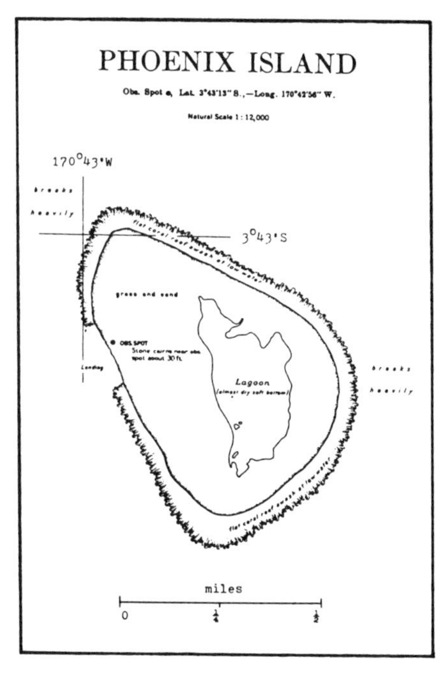

Phoenix

11. Howland and Baker

Howland and Baker are 2 small coral islands that lie about 35 miles apart, considerably northward of the Phoenix group, at 0° latitude, 176° west, about 1900 miles southwest of Honolulu. They were regarded as barren little outposts of no value, except for considerable deposits of guano, until the aviation era opened in the 1930's. Then the US asserted a claim to both, based on their having been named in the Guano Act of 1856, and token American "colonists" were landed on both in 1935. Previously they had been shown on the charts as British, but Britain made no objection to the occupation, and they are now officially regarded as American.

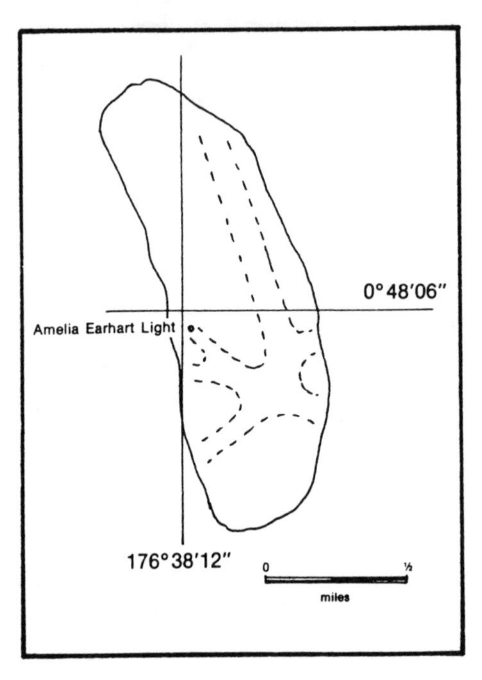

Howland

a. ***Howland.*** Howland is 1½ miles long by ½ mile wide. It is a low, flat, sandy island, covered with a moderately heavy growth of green vegetation about one foot high, with a patch of 15 foot high trees in the depressed central area. Birds, reptiles and crustaceans make up most of the animal life. It is surrounded by a fringing reef.

Howland was discovered about 1822 by the Nantucket whaleship *Oeno* under Captain George B. Worth. It was named for the lookout who first saw it from the New Bedford whaler *Isabella* in 1842. From about 1858 until 1890, American and British interests worked the guano deposits. In 1937, US authorities hurriedly built a good airfield on Howland to serve as a refueling stop for Amelia Earhart and Fred J. Noonan, who were attempting a flight around the world. The two, flying west to east, left Lae, New Guinea, for Howland on July 2, 1937, and were never seen again. The mystery of their disappearance has never been solved. The beacon on Howland is now known as "Amelia Earhart Light." The airstrip is no longer serviceable.

b. *Baker*. Baker is one mile long, 1,500 yards wide, and covers about 380 acres. It is about 25 feet above sea level and there are no trees on it. The island is quite barren except for birds, reptiles, and crustaceans. Captain Michael Baker, an American, found Baker in 1832. American interests worked the rich guano deposits in the 1850's, but this industry gradually faded out. For 20 or 30 years, no country took any interest in Howland or Baker.

In the 1880's or 90's, their deposits were worked by some British companies. In 1934 the islands were claimed by the U.S. It referred to the Act of Congress of August 18, 1856, which listed Baker and Howland among the guano islands in the possession of the US. To make American claims doubly sure, 4 Americans of Hawaiian blood were landed as "settlers" on each island in May, 1935. Following sea and air attacks by Japanese armed forces in January, 1942, the American colonists were evacuated from Baker's settlement at Mayerton and from Howland Island. The islands were regained by the US in 1944. They have been unoccupied since World War II. They are now under the control of the US Department of the Interior as part of the National Wildlife Refuge System and are visited annually by the US Coast Guard.

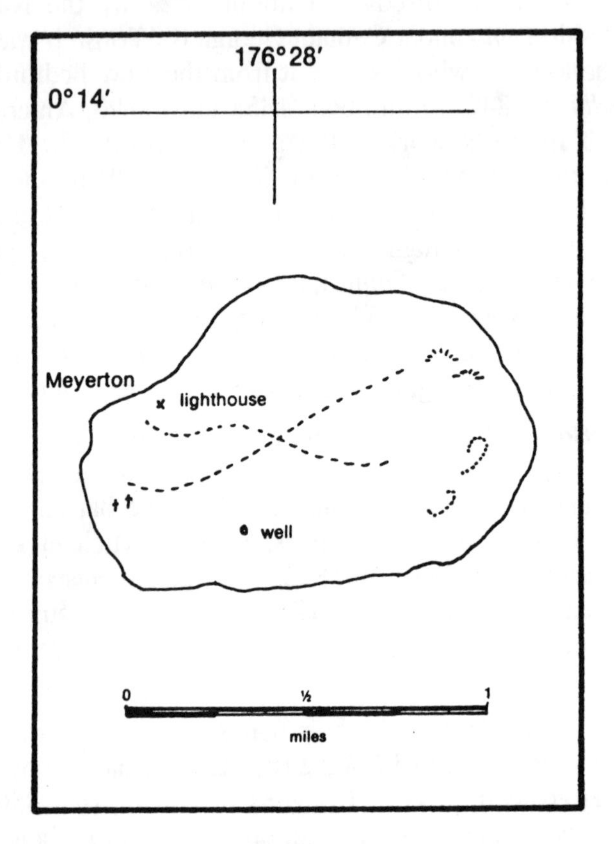

Baker

12. New Caledonia

New Caledonia is a French Overseas Territory consisting of New Caledonia, Isle of Pines, Loyalty Island and these 5 uninhabited islands and island groups:

a. ***Huon Islands.*** Huon Islands consist of 4 coral islands located about 170 miles northwest of New Caledonia at 18° south, 163° east. The islands are each about ½ mile in diameter, and together they have an area of 160 acres. They are somewhat barren and have no indigenous population, although they have been worked for guano. The 4 islands of the group are named: North Huon, Leleizour, Fabre and Surprise.

b. ***Chesterfield Islands.*** Chesterfield Islands consist of 11 coral islands lying 340 miles northwest of New Caledonia at 20° south, 158° east. The islets vary in size from ½ mile to 1½ miles long, and their areas total about 250 acres. They are well-wooded and low lying and contain rich deposits of guano, but there are no indigenous inhabitants. There has been no exploitation of the guano since 1921.

c. ***Walpole.*** Walpole is a raised limestone formation, almost flat, 230 feet high, lying 80 miles east of the Isle of Pines, at 23° south, 169° east. It is 2¼ miles long by ¼ mile wide, and about 340 acres in area. Extensive deposits of low grade guano were exported for many years. In 1920 the Austral Guano Company established moorings for ships and exported 10,000 tons of guano per year for a time, mostly to New Zealand. Due to competition from other sources, shipments declined in the 1930's and ceased entirely in 1940.

d. ***Matthew.*** Matthew lies 135 miles east of Walpole at 22° south, 171° east. It is an intermittently active volcano whose height and size vary considerably.

e. ***Hunter.*** Hunter, also known as Mearn, is another intermittently active volcano lying 43 miles east of Matthew at 22° south, 172° east. It was 974 feet high in 1958. Though frequently fuming, it has been more stable than Matthew in recent years. It is less than one mile across and has some vegetation, but it is rugged, inhospitable and virtually uninhabitable.

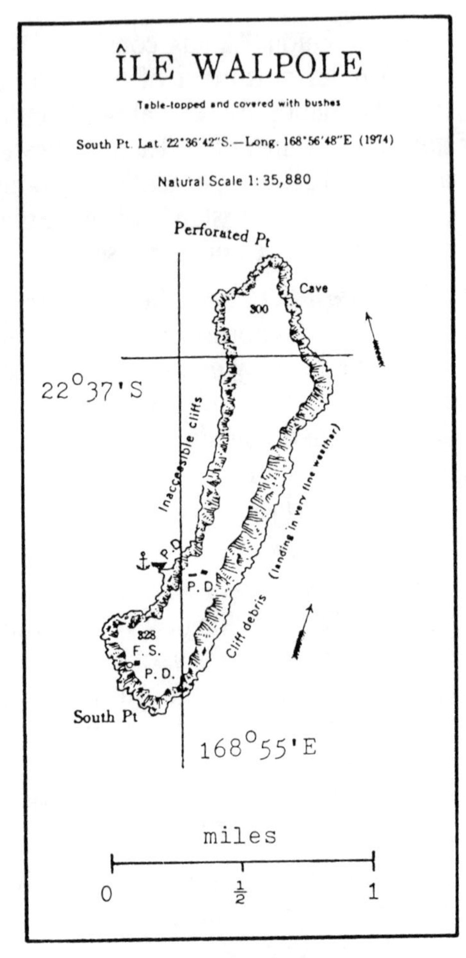
ÎLE WALPOLE
Table-topped and covered with bushes
South Pt. Lat. 22°36'42"S.—Long. 168°56'48"E (1974)
Natural Scale 1: 35,880
Perforated Pt
Cave
300
22°37'S
Inaccessible cliffs
(landing in very fine weather)
P.D.
P. D.
328
F. S.
P. D.
Cliff debris
South Pt
168°55'E
miles
0
½
1

Walpole

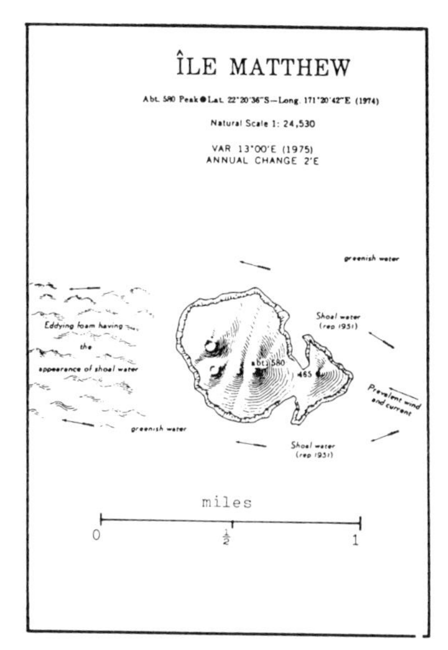

Matthew

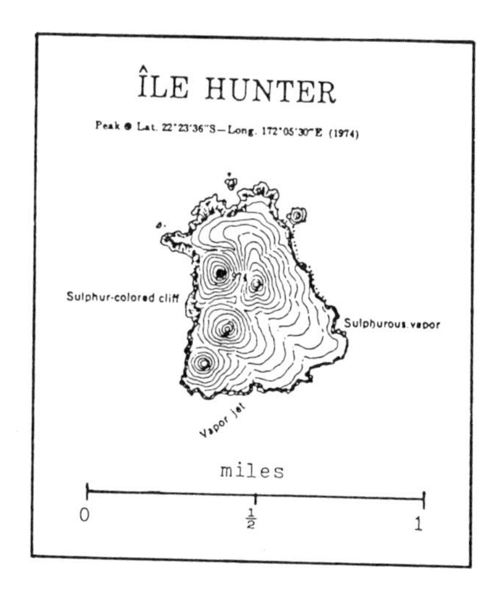

Hunter

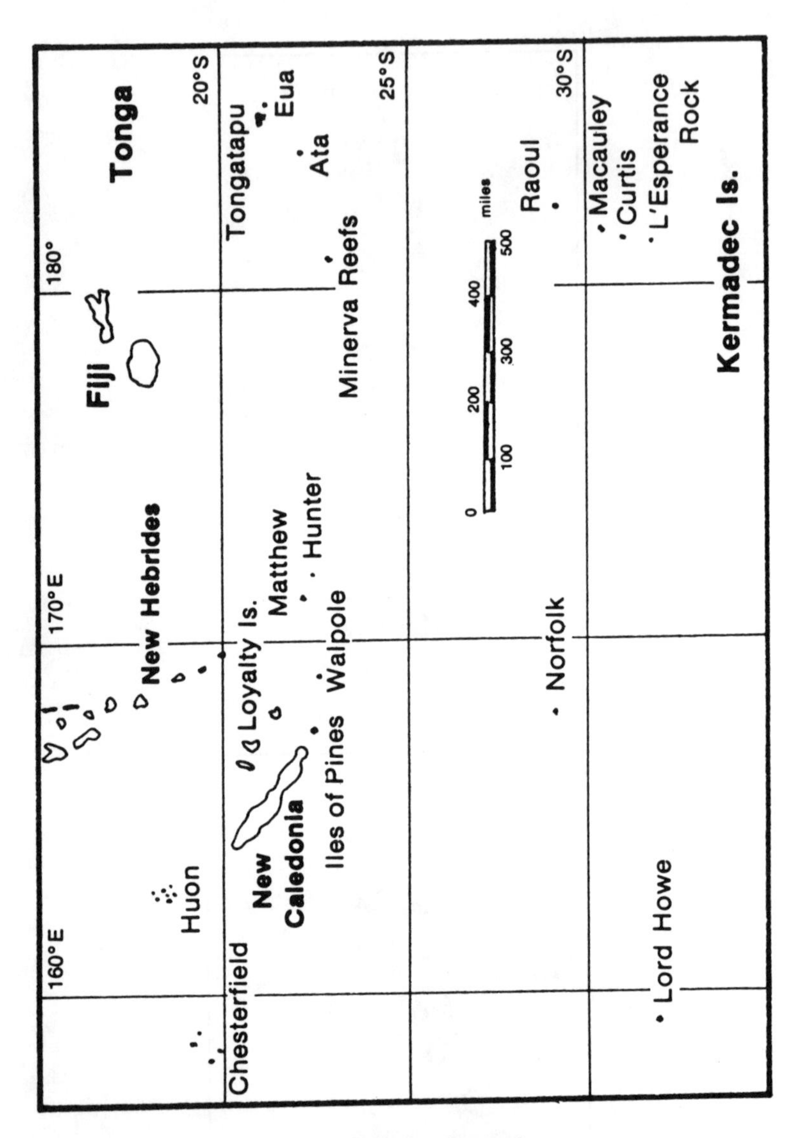

New Caledonia and Kermadec Islands

13. Kermadec Islands

At 30° south, 178° west, some 600 miles northeast of the city of Auckland, New Zealand, the Kermadec Islands stand on the inner edge of the Kermadec Trench, which runs northeast from North Island, New Zealand, towards Tonga. Frequent and sometimes severe earthquakes occur in this area. These islands are mountainous, fertile and forested. The climate is mild and pleasant with plentiful rainfall. The prevailing wind is easterly and light. Temperature varies annually from about 82° F to 48° F. Rainfall is about 57 inches per year.

The largest island of the group is Raoul (or Sunday) Island (7,260 acres), of volcanic origin with a large crater occupying much of its area. Though the highest peak is only 1,760 feet, its surface is broken by deep ravines and rocky spurs that end at the sea in steep cliffs. Northwest of it is a little group of the Herald Islets, 7 in all, with a total area of 85 acres, and of similar volcanic origin. A little to the south are Macauley (764 acres), Curtis (147 acres) and L'Esperance Rock (12 acres). These, too, are of volcanic origin with evidence of recent activity on Curtis Island.

The group was discovered by the British Navy in the 1870's. It has never had a native population. The first settlers (Baker and Reid) landed in 1837 and lived by growing potatoes and subtropical crops for sale to visiting whalers. The islands were abandoned on account of volcanic action in 1872, but in 1878 settlement started again with the arrival of the Bell family. The islands were formally annexed by New Zealand in 1886. The last of the Bell family was evacuated on the outbreak of World War I in 1914. Since World War I, settlement has been sporadic and unsuccessful, mainly due to the isolation of the group.

Before and during World War II, the New Zealand Government took a new interest in setting up a military outpost, and later some Cook Islanders were settled there to grow oranges.

At present, a weather and radio station is maintained on Raoul (Sunday) Island. The total population of this, the only inhabited member of the group, including the official staff of the station, numbers only 10. Supply ships stop at Raoul 3 or 4 times a year and radio contact is maintained with New Zealand.

Besides the isolation and the possibility of earthquake or volcanic activity, another difficulty preventing settlement of the Kermadecs is that there are no sheltered anchorages and a ship's boat can't land in rough weather. But Pitcairn Island, which is similar to the Kermadecs, also has no good anchorage, and yet Pitcairn has been inhabited for about 200 years. One way to overcome the lack of good harbors would be to use an airship (blimp), which wouldn't need a runway to land but could use any cleared meadow or beach.

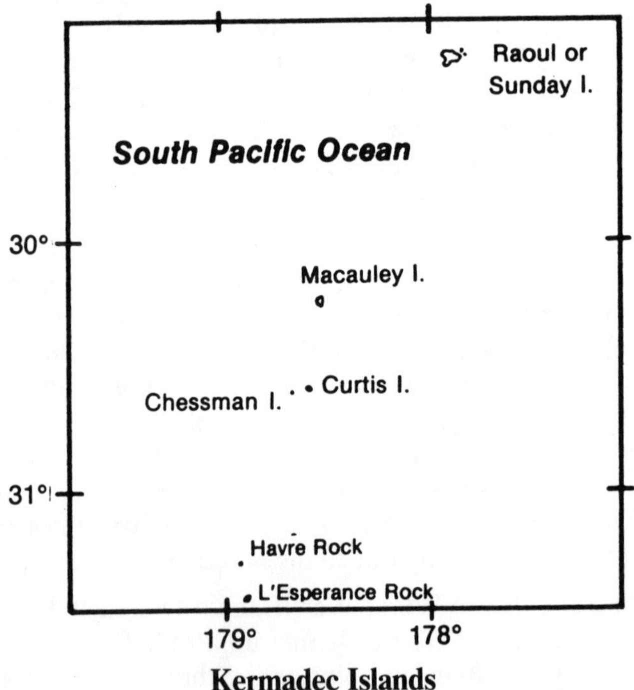

Kermadec Islands

14. Ata

Ata is an uninhabited island in the south of the Tonga Islands, 85 miles southwest of Tongatabu, at 22° south, 176° west. The island has twin peaks that arise to 1165 feet and are extinct volcanoes. There are deposits of guano suitable for manure, but lack of a harbor renders them commercially worthless. The island was formerly inhabited. In the 1860's, Ata was raided several times by slavers seeking laborers for mines in Chile. Finally, the King of Tonga, George Tupou I, ordered all 200 inhabitants of Ata to move to Eua where he could more easily protect them. Ata has been ininhabited since.

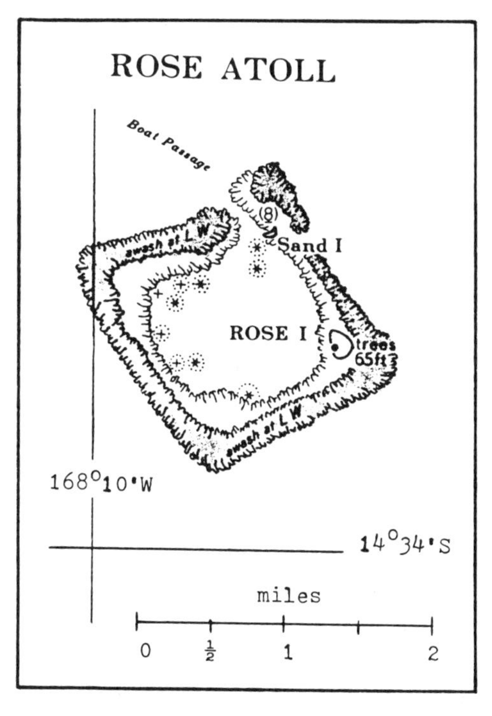

Rose

15. Rose

Rose is the most easterly island of American Samoa, lying 80 miles east of the populated island of Ta'u, at 15° south, 168° west. Rose is a small, isolated double island, a typical atoll covered sparsely with strand vegetation incapable of supporting human life. Moreover, the island is sometimes awash in heavy seas, causing such devastation that settlement has not been attempted. Vegetation is limited to coconut palms, pandamus (screw pine), casuarina trees and the usual strand growth common to the low, sandy atolls of the Pacific. Rose was claimed by the US in 1921 and is now a protected National Wildlife Refuge.

16. Suvarov Atoll

Suvarov, also know as Suwarrow, is a coral atoll at 13° south, 163° west, one of the Cook Islands, an Associated State of New Zealand. Suvarov has usually been uninhabited in the past, but it no longer is.

A New Zealander named Tom Neale who had been a storekeeper on Rarotonga for many years, fulfilled his life ambition by living in solitude on Suvarov from October 1952 to June 1954 and then again from April 1960 to December 1963. He wrote a book, *An Island to Myself,* about his experiences on Suvarov. Tom Neale died of cancer on November 30, 1977, at the age of 75.

Over the years, many other persons have visited Suvarov briefly. Since it lies in the direct line from Bora Bora to American Samoa, it has become a popular stopping place for yachts. In 1986, the Cook Islands Government appointed official residents to live on the atoll and watch over the place.

They are in daily radio contact with Rarotonga to call in weather reports and any other news. An airstrip is also being built at Suvarov to tie this once-isolated island even more closely to the civilized world.

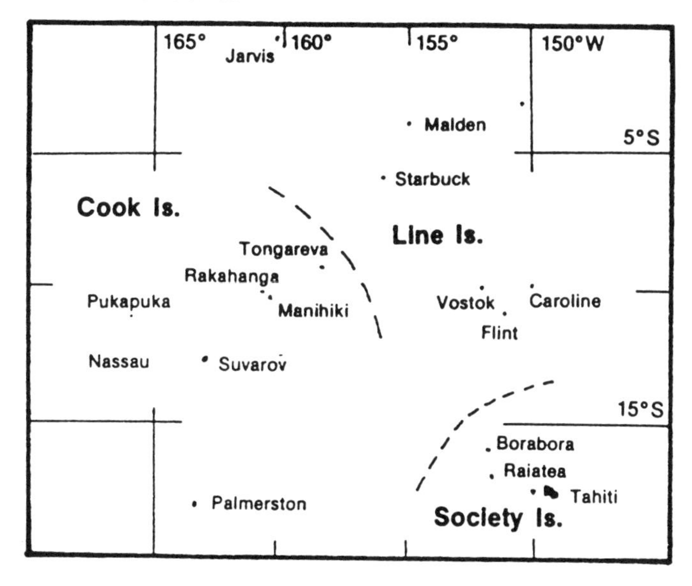

Cook Islands and Line Islands

17. Tahiti

Tahiti is the largest and most important island of French Polynesia, and its capital of Papeete is the administrative center of all the French islands in the eastern Pacific. Tahiti has a population of about 96,000. These uninhabited islands lie near Tahiti:

a. *Mehetia.* Mehetia, also known as Maitea, is an uninhabited island of volcanic origin lying 60 miles east of the southern tip of Tahiti at 18° south, 148° west. Mehetia is a small, circular island with an area of 600 acres. Very high and steep, it rises to 1,427 feet, and is surrounded by a fringing reef. It is visited occasionally by fishermen from the south side of Tahiti, but they seldom go ashore, since it has no good landing place. There is one spot where it's just a little less dangerous to attempt to get ashore than at other places. Even here, the attempt should only be made at high tide when the sea is calm.

Mehetia once had a population of 100, but late in the 19th century the French authorities moved the last remaining Polynesians to other islands. Then a certain Austrian heard about the island and decided to buy it. To do that, he had to clear the title by contacting the 100 or so families who claimed ownership, who were by then scattered all over French Polynesia. He died before he could complete this epic feat. The task was taken over by his nephew, Marcel Krainer, who came out to Tahiti from Austria in 1924 and spent another 10 years tracking down Mehetians. Finally, by the early 1930's, Krainer convinced the French Government that he had all necessary approvals, and after he paid the Government about $100, they declared him the owner of Mehetia.

After completing the purchase of the island, Krainer tried raising pigs on it for profit. But the venture went bust. Landing on the island, and loading and unloading supplies, proved to be just too hard. Eventually, he sold the island. The present owner is thought to live in Papeete, Tahiti. No one lives on Mehetia.

b. *Duke of Gloucester Islands.* Duke of Gloucester Islands, also known as Anou Islands, or Iles de Duc de Gloucester, are an uninhabited group lying about 400 miles southeast of Tahiti at 21° south, 143° west. The group consists of three atolls: Anuanu Raro, Anuanu Runga and Nuku Tipipi. They are heavily wooded with a heavy surf and no sign of an anchorage. The middle atoll, Anunanu Runga, is about 5 miles long.

18. Tubuai Islands

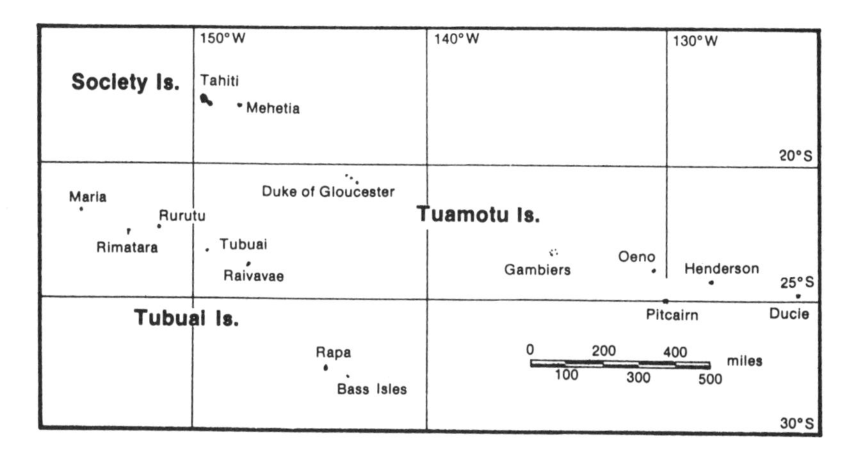

Tahiti, Tubuai Islands and Pitcairn

Tubuai (or Austral) Islands include 5 inhabited and 2 uninhabited islands forming a chain extending southeast to northwest over 800 miles. The nearest of these to Tahiti is Rurutu which lies about 300 miles south of Tahiti. The Tubuais are part of French Polynesia. The 2 uninhabited Tubuais are Bass Rocks and Maria.

a. *Bass Rocks.* Bass Rocks (or Isles), also known as Marotiri, consists of 4 tiny islands at 28° south, 143° west, 46 miles southeast of Rapa, at the southeasterly end of the Tubuai

Islands. Bass rises to 346 feet and is visible from Rapa in clear weather. There are no good landing places.

b. *Maria.* Maria, also known as Hull, lies at the opposite northwesterly end of the Tubuai Islands at 22° south, 154° west. It is an uninhabited atoll consisting of 4 islets. Maria is about 120 miles northwest of Rimatara and it is occasionally visited by natives from there on fishing or copra cutting expeditions.

19. Marquesas Islands

The Marquesas Islands, a remote backwater of French Polynesia, lie in the southeast corner of the Pacific. They consist of 13 small islands, only the 6 largest of which are inhabited. The inhabited Marquesas are the northern group of Ua Pou, Ua Huka and Nuku Hiva, which is the administrative center and the port of entry for the Marquesas. The southern group consists of Fatu Hiva, Tahu Ata and Hiva Oa, which is the main island of the southern group.

All these islands are of volcanic origin with steep mountains dropping abruptly into the sea, and no protective barrier reefs. The climate is warm and humid at around 10° south latitude, but there are occasional periods of drought. The vegetation is mostly thick tropical jungle. The mountains and central plateau of several islands rise to several thousand feet and have a cooler, subtropical climate as a result. The plateau is open and grassy with some scrubby bushes. These islands have no native large animals, but horses, cattle, sheep, pigs, dogs and other domestic animals gone wild roam the plateau. The natives live mostly in a few narrow, fertile coastal valleys. None live at the higher elevations where the nights are too cool for them. Marquesan crops include breadfruit, yams, coconuts, mangos, coffee and vanilla. Copra is exported.

The native population of the Marquesas, which was over 50,000 in the 18th century, declined to a low of about 1,300 in 1936 as wars and western diseases took a devastating toll. Since then the population has risen again to a reported 5,588 by 1963. Several formerly inhabited islands remain completely depopulated. The main islands are now inhabited by a handful of white planters and a few thousand Polynesians, nearly all of whom live in narrow valleys near the coasts.

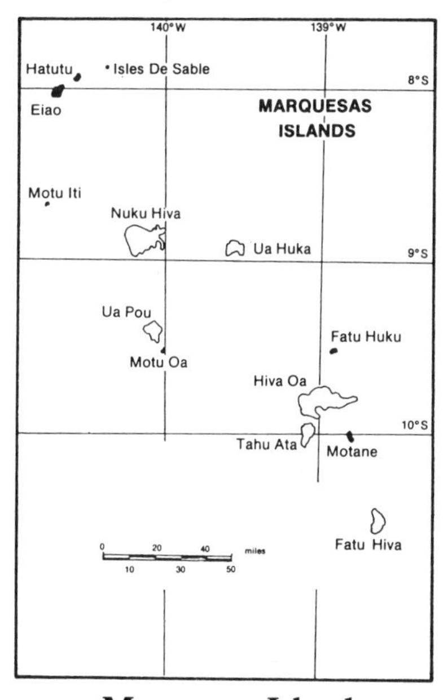

Marquesas Islands

Presently, the Marquesas are much less isolated than formerly. For example, Marquesas Islanders didn't hear about the outbreak of World War I until the war had ended. But these days ships make regular stops, at least at Nuku Hiva, and there are now 2 or 3 small airfields in the Marquesas, and small planes make regularly scheduled flights between the Marquesas and Tahiti. Despite the underpopulation of these islands, the French authorities remain cool to westerners who might want to settle here, but they welcome tourists willing to spend large amounts of money.

Here are descriptions of all the uninhabited Marquesas:

a. *Fatu Huku.* Fatu Huku (Hood Island) is about 15 miles north of Hiva Oa at 9° south, 139° west. This uninhabited islet is a rock that rises to 1,180 feet high. It contains some interesting burial caves and is visited occasionally by fishermen and bird hunters. Very few white men have ever been on the island.

b. *Motu Oa.* Motu Oa is a small, flat-topped rock islet about 400 feet high, off the south coast of Ua Pou. It is covered with grass and millions of sea birds nest here.

c. *Motu Iti.* Motu Iti is a flat-topped rock islet located northwest of Nuku Hiva at 9° south, 141° west. At its highest point it reaches 720 feet.

d. *"The Bird Island."* In his book, Danielsson tells about a visit to an island that he identifies only as "the bird island," where he stopped while sailing from Ua Pou to Ua Huka. Danielsson says (page 158): "The bird island proved to be a low, flat rock which rose about 60 feet above the surface of the sea. The rock, which was certainly 500 yards long and almost as wide, was as flat as a floor and appeared to be quite inaccessible. We went round it for some distance, but everywhere I saw the same high, perpendicular wall, rising straight up out of the rough sea."

With some difficulty, Danielsson's party did succeeed in climbing the cliff and they found the top covered with nesting sea birds. They easily gathered a boatload of eggs and birds for a feast. Later they visited a lobster cave on the north side of the bird island, where the sailors quickly speared a few dozen large rock lobsters.

Plate 25, facing page 177 in Danielsson's book (see references) has 2 photos of "the bird island," which appears completely barren and treeless.

e. *Motane.* Motane (sometimes called Mohotani) is a small island 5 miles by 2 miles, located 11 miles south of Hiva Oa's eastern end, at 10° south, 139° west. It rises to 1,706 feet at its highest point. Motane once had a native population, but it has been uninhabited now for over 100 years. It is steep, rocky, dry and overrun with sheep. Schooners sometimes stop there so the sailors can hunt the sheep.

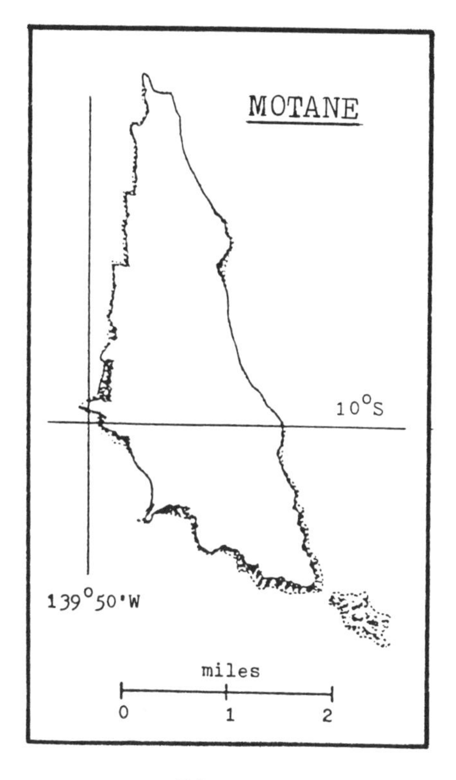

Motane

Motane has been the victim of an ecological disaster. When the Spaniards passed by in 1595, they described it as "an island beautiful to look at, with much wood and fair fields" (Heyerdahl, page 220). But by the time Heyerdahl visited it in 1936, Motane had become a completely treeless, arid desert.

Describing his visit, Heyerdahl notes the hot, white, dry ground made up of sterile, sandy soil and bare rocks. Here and there he saw some dry shrubs, but no trees, and no green vegetation. Because the canopy of trees was gone, the tropical sun beat down mercilessly on the dry ground. The only remnants of the once-abundant forests were a few dead white trees, like dry bones, without either leaves or bark.

There were plenty of actual bones scattered all around, too, and even complete skeletons of sheep. He saw a number of stone house platforms which are used in the Marquesas to raise houses up out of jungle mud. But on Motane there was nothing left to make mud, not much soil, and especially not a drop of water. Dried up streambeds indicated that there had once been running water on the island, however.

The only signs of animal life Heyerdahl observed were the wild cock he heard crowing, and a few scrawny sheep he saw running through some dried-up bushes.

Heyerdahl speculated that the death of Motane began when the last human inhabitant died or sailed off to another island. Then the livestock ran wild with no carnivorous animals to keep them in check. The sheep flourished, for a time, and there was a population explosion among them, increasing their number far beyond the capacity of the island to feed them. So hordes of wild sheep ate everything: grass, the roots of the grass, all the leaves they could reach, and finally the bark of the trees. Stripping off the bark killed the trees. Without the trees, the sun baked the ground dry, and without roots to hold the soil, when tropical storms hit, most of the soil was washed away. The sheep population collapsed, leaving only their bones behind. That, Heyerdahl thought, was how Motane turned into the desert that he found it to be. (Heyerdahl, page 185, see references.)

f. *Terihi.* Terihi lies just off the southeast coast of Motane and rises to 804 feet.

g. *Eiao.* Eiao (Masse Island) lies 56 miles northwest of Nuku Hiva at 8° south, 141° west. It is about 6 miles long by 3 miles wide, and it rises to 1,889 feet. There is a good landing place at a bay known as Vai Tahu on the northwest corner. There are ruins of a Polynesian population from the distant past. More recently the French used the island as the site of a prison. Domestic animals were introduced during these earlier occupations, and sheep, cattle, and pigs have now gone wild and overrun the island. There are also herds of wild asses of enormous size..

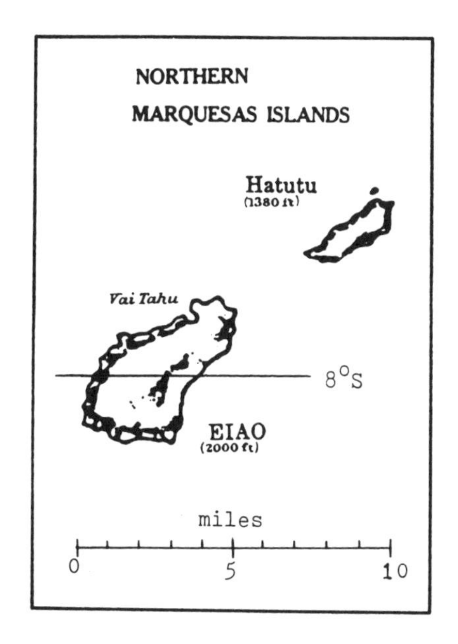

Northern Marquesas

h. *Hatutu.* Hatutu is the most northerly of the Marquesas, lying 10 miles northeast of Eiao. Hatutu is about 4 miles long by one mile wide and rises to 1,404 feet. It is home to thousands of ground doves not found on any other Pacific island. There is no water on the island, but there are signs of former habitation.

i. *Iles de Sable.* The dots on some maps north of Hatutu, which are named Iles de Sable, are merely sand banks, perhaps 500 yards long by 100 yards wide, and only about 10 feet high. The sea in that area swarms with enormous sharks.

20. Pitcairn Group

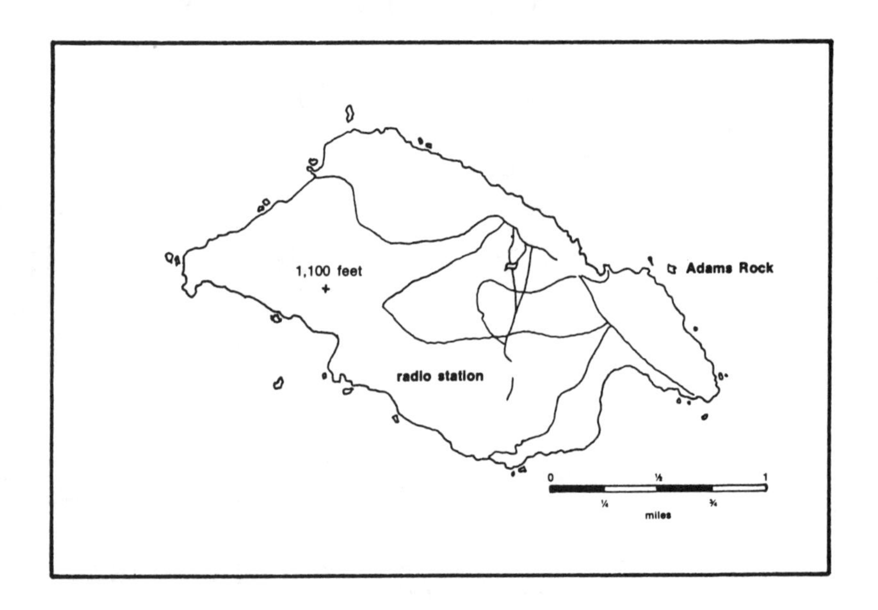

Pitcairn

The Pitcairn Group includes the populated island of Pitcairn (1.8 square miles) and the nearly uninhabited islands of Oeno, Henderson and Ducie. Pitcairn lies isolated in the extreme southeast corner of the Pacific Ocean with only lonely Easter

Island lying farther east. Pitcairn is inhabited by people of British/Tahitian descent whose ancestors were the *Bounty* mutineers and their Polynesian wives. Pitcairners were converted to Seventh Day Adventism long ago, and most of them are very religious. The population of Pitcairn has been as high as 233 in 1937, but in recent decades, as communications with the outside world have increased, and as Pitcairners have become more aware of opportunities elsewhere, there has been an increase in emigration, particularly to New Zealand. As a result, the population has been declining since the 1930's, and by 1982 it was down to about 60, including only 7 children. There is a real possibility that the island may become completely depopulated. Pitcairn has long been a self-governing British Colony. Presently it is administered by a British representative in New Zealand and a local council on Pitcairn.

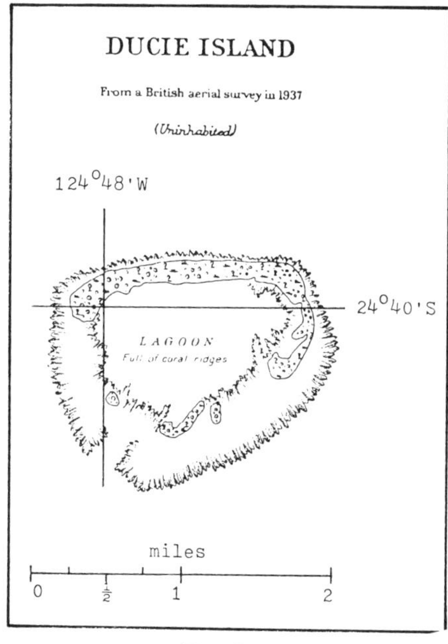

Ducie

Oeno, Henderson and Ducie are sometimes visited by parties from Pitcairn — otherwise, they are uninhabited. Oeno lies 65 miles northwest of Pitcairn at 24° south, 131° west. Henderson is 100 miles to the northeast of Pitcairn at 24° south, 128° west; and Ducie is 190 miles east of Henderson at 25° south, 125° west. Oeno and Ducie are small coral atolls with sheltered lagoons. Henderson is as larger island that consists of an unusual raised coral formation, one of only a small number of high coral islands found in the Pacific. All 3 islands were made part of the British Pitcairn Colony in 1938.

Oeno

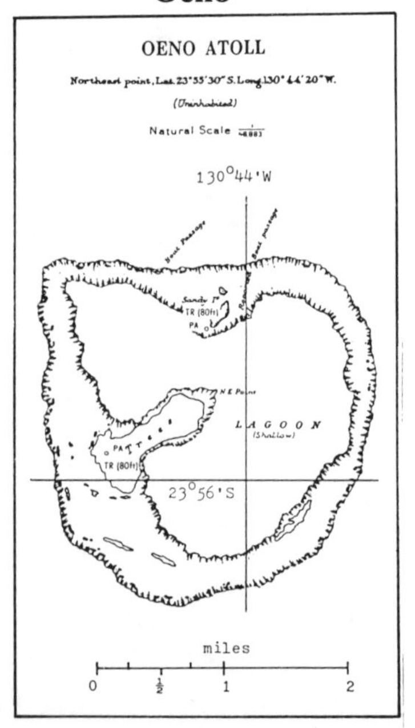

Oeno is flat throughout, rising only a few feet above sea level. Coconuts and a few orange and lemon trees grow there, and inland there is a hollow from which fresh water may be obtained if a hole is dug in the sand. Oeno is sometimes visited for a few days a year by Pitcairners on an expedition which is part picnic and camping trip, and part business. They do some fishing and take the dried and salted fish back home. They also gather coral shells and pandanus leaves that are used in making curios, which is Pitcairn's main source of cash.

Henderson

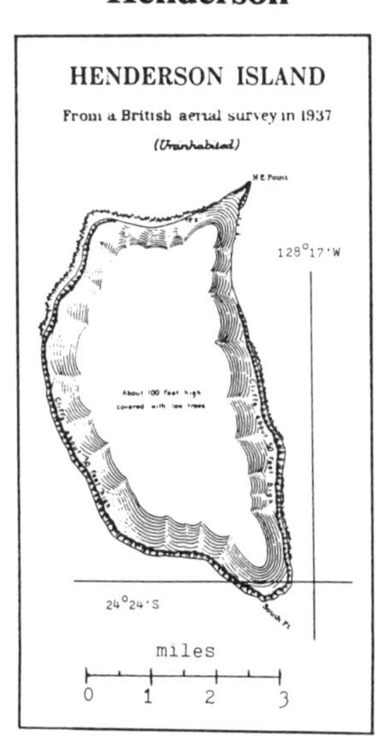

Henderson, formerly known as Elizabeth, is 5 miles long by 2½ miles wide, and up to 100 feet high. At about 12 square miles in area, uninhabited Henderson is much larger than the nearby populated Pitcairn, which has less than 2 square miles of area. Henderson is ringed by 50 foot high cliffs except for a stretch of coast on the north and east. Inland, the terrain is covered with low trees. In the past, Pitcairners used to visit Henderson to gather hard, durable wood known as miro which they used in curio-making. But the trip is hazardous across 100 miles of ocean in small, open longboats which are the only vessels Pitcairners have. Years ago, they sometimes were given a lift to

Henderson on ocean liners, but few big ships pass near Pitcairn anymore.

Henderson has no natural harbor, and fresh water is not available on the island because high coral islands are very porous and rainwater runs right through. A settler could solve the water problem by constructing a cistern to hold rainwater, or by desalinating seawater.

In recent years Henderson Island has been in the news because of the interest shown in it by an American who considers it his ideal uninhabited island. Arthur M. Ratliff, a multi-millionaire farmer and businessman from Virginia, visited Pitcairn in 1981 in a chartered yacht and found it to be his ideal place to retire and get away from it all. Pitcairn itself was a little too small, but about 100 miles away lay Henderson whose empty 12 square miles seemed broad enough for Ratliff's ambitions. So Ratliff made the British Government an offer: in return for the title to Henderson Island, Ratliff would donate $800,000 to the Pitcairn Island development fund for the benefit of the 50 or so remaining Pitcairn Islanders, and he would build a small airstrip on Pitcairn and a larger one on Henderson, and he would give the Pitcairners 3 small airplanes, and he would allow the Pitcairners to come along on the flights of his own family in and out of Henderson Island. That would end the isolation of Pitcairn and give the Islanders easy access to the outside world for medical treatment or to bring in supplies. The Pitcairns wouldn't even have to put up with their benefactor in their midst because he would be on another island 100 miles away.

This was the best offer the Pitcairn Islanders would ever receive, but opposition to it in Britain was immediate and intense. What will happen to the wildlife on Henderson Island, the critics demanded to know. There are unique species of birds and flowers and snails on Henderson that are found nowhere else, they said, which would be harmed or destroyed if Ratliff were allowed to proceed with his plans. Of course, every isolated island has unique species. That's characteristic of places where wildlife evolves in isolation. Even so, it often takes an expert to tell the difference between these unique species, and the similar common varieties found elsewhere.

One is led to suspect, however, that the opposition to Ratliff's proposal was fueled not so much by concern for Henderson's plants and animals, as by the character and background of Ratliff himself which British snobs found highly offensive. Ratliff, it seems, is a man of humble and poor origin, with no pedigree, who made his fortune in coal mining; and worst of all, he's an American. One British magazine, fuming with indignation, described Ratliff as "a tobacco-chewing, hillbilly millionaire from the backwoods of Virginia." No doubt if Ratliff had been a British Milord, with a proper British accent and a centuries old pedigree, going back to the bastard son of some Elizabethan pirate perhaps, why then Brits wouldn't much give a damn about Henderson's snails. Of course, these days it's getting hard to find a British Milord worth $100 million.

As might have been predicted, as a result of this widespread anti-American opposition, the British Government turned down Ratliff's proposal. There was some talk about the British Government itself providing a couple million dollars to aid the Pitcairners, but it would be politically difficult to justify giving some $2 million to 50 people on Pitcairn, without also giving $2 million to every 50 destitute people living in the slums of British cities, so nothing came of the proposal, and the Pitcairners have been left where they were before Ratliff's offer: slowly sinking.

The future of Pitcairn doesn't look promising. The population, now around 50, keeps falling as Pitcairners move away, often to New Zealand. They can now just barely muster enough able-bodied men to man the boats that are the only means of getting on or off the island. It seems likely this migration will continue until Pitcairn is entirely abandoned and itself becomes as deserted as Henderson, Oeno, and Ducie already are.

21. Sala Y Gomez

Sala y Gomez is an arid, barren, uninhabited rock, lying 210 miles northeast of Easter Island, at 26° south, 105° west. Along with Easter Island, Sala belongs to Chile, which lies 2,100 miles to the east. Sala is about 3,900 feet long by 500 feet wide, giving it an area of about 44 acres.

22. Clipperton

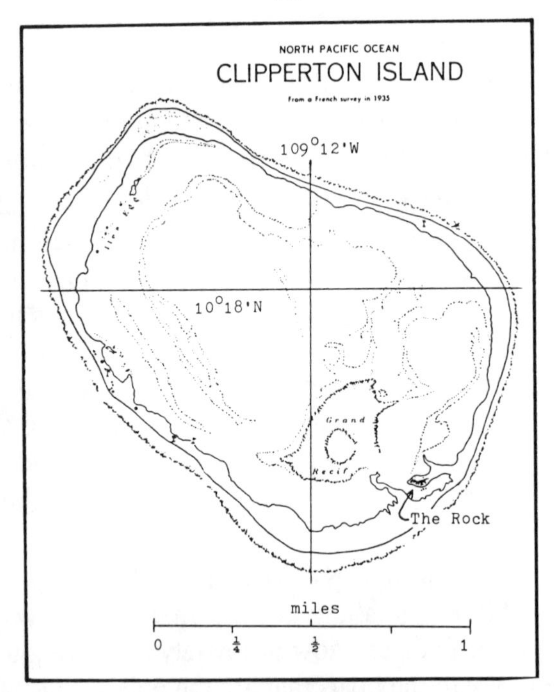

Clipperton

Clipperton, which lies southwest of Mexico at 10° north, 109° west, is an oval coral atoll, about 2 square miles in area. It is about 1,800 miles due west of the Panama Canal, and is politically part of French Polynesia, even though it lies over 2,500 miles from the nearest French islands (the Marquesas).

Clipperton is surrounded by a reef about 5 miles in circumference and has a fresh water lagoon with no outlet to the sea. A coral shelf extends for a considerable distance out from the shore, making an almost impassible barrier for boats. So the island is difficult and dangerous to approach by sea, and there is no safe anchorage.

Most of the island is a low sandbank, barely 10 feet high, except for a jagged promontory on the southeast side known as "The Rock," which rises to 70 feet. Vast flocks of seabirds such as terns, boobies and frigates nest on the island in certain seasons. Clipperton used to be covered with vegetation, but now it is overrun by crabs, which have denuded the atoll of most of the vegetation except for the few established trees, which are mainly concentrated into 2 small coconut groves. Since the crabs eat any coconuts that fall to the ground and sprout, no new coconut palms can get started.

The island had a more balanced ecology years ago when pigs ran wild. The pigs arrived on the ship *Kinkora* which ran aground on Clipperton's reef in 1897. A couple of pigs survived the shipwreck and began preying on seabird eggs, chicks, sick birds, land crabs, and whatever food they could root up. By 1958, Clipperton's swine herd had increased to 58. Scientists who were on the island conducting research that year noticed that the lush vegetation contrasted with old descriptions of the place as a barren atoll. Land crabs no longer swarmed in huge numbers as they had done around 1900. And the seabirds seemed to have declined, with only hundreds nesting compared to the thousands mentioned in old accounts. The scientists attributed these changes to the feral pigs, and to save the birds, they set out to shoot them all, and succeeded in completely liquidating Clipperton's swine population. Ironically, they didn't realize it was not yet the breeding season for the seabirds, which probably accounted for their low numbers. Earlier visitors had reported that the birds seemed well able to drive off any pigs that came too close to their nesting areas by assaulting them from the air. Anyway, by 1980 when Jacques Cousteau led an expedition to Clipperton, it was once again a barren place, overrun by land crabs, and the crabs were seen to be preying heavily on nesting seabirds. While it's generally true that domestic animals gone wild often harm the ecology of isolated islands, Clipperton seems to be an exception in that its ecology is sounder with pigs than without them.

The sea surrounding Clipperton swarms with fish, sharks, and moray eels, but only tiny crustaceans, bacteria, and weeds live in the now-landlocked lagoon. Once the lagoon was open to the sea and sea creatures lived there as well, including reef-building corals. But as the reef grew, the opening to the sea slowly shrank, until it had closed completely by about the middle of the 19th century. Up to 200 inches of rain falls annually on Clipperton, so once the lagoon was sealed off from the sea, the rain gradually converted the lagoon into a fresh water lake, killing all the salt water creatures that were trapped there, including the reef-building corals. Now the water in the lagoon is said to be brackish but potable, that is, a little salty but drinkable.

There are several small islets in the lagoon. Skin divers who explored the deepest part of the lagoon encountered an unexpected burning sensation in their eyes and on their skin. Later tests revealed that there is a high concentration of hydrogen sulfide at the bottom of this deep hole, apparently produced by dead vegetation that falls to the bottom and decomposes.

Clipperton was discovered by Spanish explorers. John Clipperton, a British privateer who preyed on Spanish shipping in the western Pacific, reportedly saw Clipperton in 1705. Legend has it that he used the island as his hideout, but that seems unlikely since it has no harbor and would be of little use to a pirate. Most likely, John Clipperton never set foot on the island that bears his name.

France claimed the island in 1858, but didn't occupy it. In the 1890's, an American company with headquarters in San Francisco mined guano (bird droppings used for fertilizer) on the island. But Clipperton Island guano was so low quality it couldn't compete with richer deposits found on other Pacific Islands. Mining operations gradually petered out, and ceased entirely in the early 1900's.

In 1897, Mexico took control of the island and kept a garrison there for many years. Mexico abandoned its garrison on the island when World War I broke out, and left the people stranded on the island to endure a 3 year struggle for survival. During those years, most of the inhabitants of Clipperton died of starvation or sickness.

In 1915, after a year had passed with no word from the mainland, a ship was sighted. When it became apparent that the ship was not going to stop at Clipperton, Captain Ramon d'Arnaud, the commander of the garrison, set out in a small boat with all the soldiers who were left, in a desperate attempt to contact the ship. Observers on shore saw them lose control of the boat in a heavy surf above the reef. A few hours later a storm rolled in and no trace of Captain d'Arnaud or his companions was ever found.

The only man remaining alive on the island was Alvarez, the lighthouse-keeper who lived near The Rock. When the storm had passed, he suddenly appeared, seized all the guns, and proclaimed himself King of Clipperton. He demanded the services of the women, and over the next 22 months, he brutally dragged off to his hut whichever of the women or girls suited his fancy at the time. Finally, in 1917, the women, led by the widow of

Captain d'Arnaud, overpowered Alvarez and killed him with a hammer. By an amazing coincidence, that very day a US Navy vessel, the *Yorktown,* arrived at the island, checking to see if Germans were using it as a base. The *Yorktown* rescued the 3 women and 8 children who were the only survivors.

After the war, France reasserted its prior claim and was awarded the island after arbitration in 1930. Mexico turned the atoll over to France in 1932. The French surveyed the island by ship and plane in 1935. The US Navy occupied the island in 1943 for the duration of World War II. During the International Geophysical Year of July 1957, to June 1958, a team of French scientists was stationed on Clipperton. Since then the island has remained mostly uninhabited, except for occasional visits by scientists.

Such a visit was made a few years ago by a film and research team led by Jacques Cousteau. They airlifted supplies onto the island by helicopter from a ship anchored offshore. And they cleared a temporary airstrip on which a fixed-wing aircraft could land to fly in supplies from the mainland. Besides conducting scientific research, the Cousteau team made a film entitled *Clipperton — The Island Time Forgot* which has been shown on the PBS television network in the US.

23. San Felix and San Ambrosio

About 600 miles west of Chile in the South Pacific Ocean, at 26° south, 80° west, lie the small uninhabited islands of San Felix and San Ambrosio, which are possessions of Chile. San Felix is about 3 miles by one mile in extent, and it rises to 600 feet. Just off the southeast tip of this arid island lies an islet known as Gonzalez. About 12 miles southeast of San Felix lies the smaller, higher island of San Ambrosio, which is about one mile in area, rises to 1,570 feet, and is equally barren.

These islands were discovered by Juan Fernandez in 1574. In 1788 the *Lady Washington,* an American sloop, visited San Felix. They managed to find a little fresh water, and they killed a large number of seals, especially on San Ambrosio, for their skins and oil. Over the years other sealers worked these islands until the seal population was wiped out.

In the 1950's and 60's, various Chilean companies that were primarily involved in fishing for lobsters at the Juan Fernandez islands also visited San Felix and San Ambrosio from time to time in search of lobsters. But the lobsters seem to be going the way the seals went, as the supply in these waters is diminishing.

24. Islands of South Chile

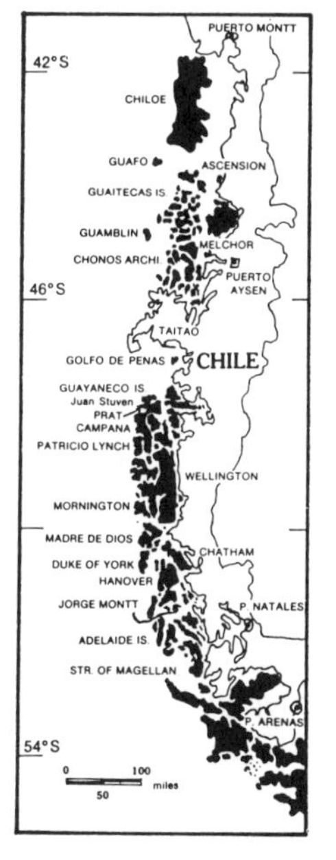

Islands of South Chile

On the west coast of South America, from the southern shore of the Island of Chiloe at 43° south latitude, all the way to Cape Horn at the tip of Tierra del Fuego at 56° south, a vast labyrinth of islands and steep-sided fiords extends for 1,000 miles. This region has been mapped only from the air and sea and otherwise remains largely unexplored, unsurveyed and unpopulated. The region from Chiloe to Puerto Natales, a town (population 11,500) at the northern limit of Tierra del Fuego, which is a distance of about 600 miles, comprises the Chilean province of Aysen. The entire province had a population of 48,000 in 1970, almost entirely living on the mainland. The provincial capital is the town of Aysen, on the mainland river of the same name, which had a population of 7,140 in 1970. Puerto Aysen is the largest settlement in the province.

One reason settlers have been slow to move into this area is because of broken terrain with steep mountains, fiords, and some glaciers that run right down to the sea, making it impossible to build a road running the length of the province. A motorist has to cross the Andes and travel north or south through Argentina. On the Chilean side, the only way to get from Chiloe and points north to places in Aysen Province or to Puerto Natales and points south is by air or by sea.

On these islands themselves the only settlements are Melinka on Ascension Island, with a population of 450, Puerto Lagunas on Melchor Island, which has a radio station, and Puerto Eden on the east side of Wellington Island, which was a weather station and seaplane base in the 1950's. Alacaloof canoe Indians once dominated this region, living on seals and shellfish, but they are now dying out. There were about 4,000 of them in 1850. By 1950 they were down to 300. And today there are only 5 Alacaloof families left, consisting of 27 men, women and children. These few survivors live at Puerto Eden on Wellington Island. The only other persons one might see along these shores in certain seasons are a few fishermen and otter and seal hunters.

The climate on these islands is cold and wet. These mountainous, misty islands are covered with dense forests of both evergreen and deciduous species. In some places, bamboo grows in impenetrable thickets. And an abundance of thorny shrubs also impedes surface travel. These woods are home to numerous birds including, among others, hummingbirds, parakeets, and woodpeckers. The mammals found here include deer, fox, opposum, puma and various rodents. Along the coasts one finds fur seal, sea otter and nutria (similar to beaver) which have valuable fur, and seabirds and shellfish.

From the island of Chiloe on south, this region includes the Chonos Archipelago which runs for 130 miles and includes thousands of islands. Below that lies Taitao Peninsula, which is almost an island since it's connected to the mainland only by the narrow Isthmus of Ofqui. The southern shore of the Taitao is washed by the Golfo de Penas (Gulf of Troubles), a turbulent bay where many ships have been lost. No outlying islands shelter this coast so the waves roar in unimpeded from the open Pacific and from Antarctica thousands of miles away.

South of this gulf one finds mostly larger islands with even fewer people than the few who live on the islands north of Taitao. This area has been described as: "a vast maze of islands and channels where there is no sign of human life to intrude upon the grandeur and desolate solitude of nature." These more southerly islands often run to several hundred square miles in area and they rise up to a couple thousand feet. The largest island in this area is Wellington, which is 100 miles long by 15 to 25 miles wide and has the highest elevation of 3,300 feet. Wellington is mountainous and has glaciers, swamps and forests. Its coastline is very irregular and deeply indented with fiords. There is a tiny settlement named Puerto Eden on Wellington's east coast, but it is otherwise uninhabited.

Addtional information about the major islands running as far south as the Strait of Magellan is given in the table below. There are still more uninhabited islands south of the Strait in the province of Tierra del Fuego, which have an even harsher sub-Antarctic climate.

MAJOR ISLANDS BETWEEN CHILOE AND THE STRAIT OF MAGELLAN

islands	*latitude*	*dimensions (miles)*	*populated places*
Guaitecas Islands			
Ascension	44°	21 sq mi	Melinka
Chonos Archipelago			
over 1,000 islands	44° to 46°		a few Indians
Melchor			Puerto Lagunas
Guamblin	45°	13 x 5	
Taitao Peninsula	46° to 47°	75 x 70	Isthmo de Ofqui
Guayaneco Islands	48°		
Byron	48°		none
Wager	48°		none
Juan Stuven	48°	20 mi long	none
Prat	48°	27 x 12	
Campana	48°	50 x 12	
Patricio Lynch	49°	27 x 3 to 10	
Wellington	49° to 50°	100 x 15 to 25	Puerto Eden
Mornington	50°	28 x 8	none
Madre de Dios	50°		none
Duke of York	51°	23 mi long	none
Chatham	51°	35 x 12	none
Hanover	51°	40 x 5 to 22	none
Jorge Montt	51°	28 x 25	none
Adelaide Islands	52°		almost entirely uninhabited

Part II
Sub-Antarctic Islands, Pacific Sector

25. The Snares

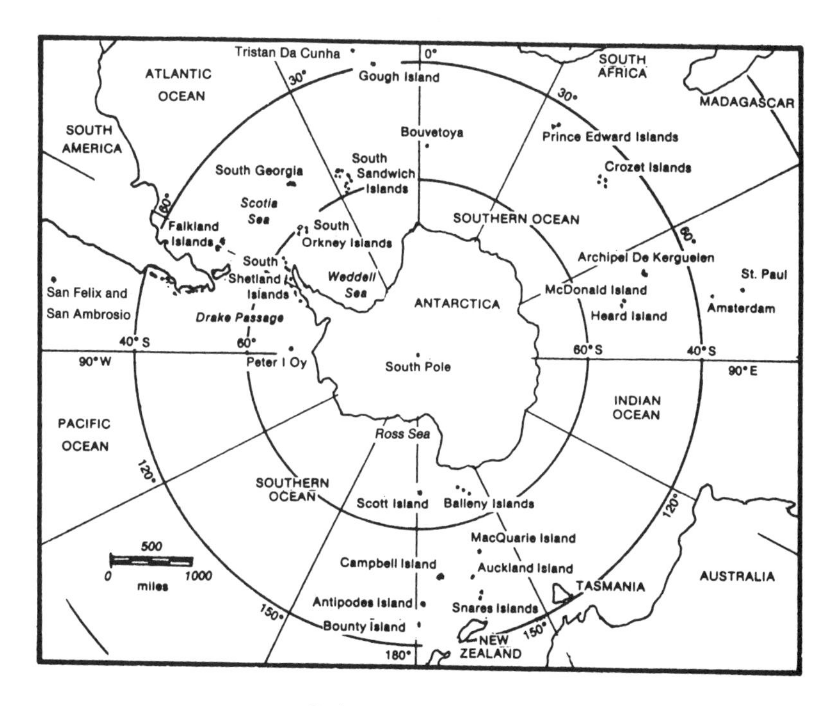

Sub-Antarctic Islands

Across the Pacific from South America, near New Zealand, lie The Snares, a group of uninhabited islets at 48° south, 167° east. They are distant outliers of Stewart Island (65 miles to the north), which is the most southerly of the New Zealand islands. The Snares consist of one large triangular island, with several offshore islets and a string of bare, rocky islets running off to the southwest along a rocky reef known as Western Reef. All of the islands in the group are made of coarse granite rocks generally similar to those of Stewart Island. The main island, whose highest point is 620 feet, is bounded by steep cliffs on the west and south, and slopes down to the northeast where there is a good anchorage. The weather is windy, cloudy, wet and cool, with the prevailing winds from the north and west. The vegetation runs to stunted forest with patches of coarse tussock grass. There are widespread beds of peat impregnated with bird dung. Sea birds in large numbers and seals are found there, but there are no land mammals, either native or introduced.

The Snares were discovered by Vancouver and Broughton independently on the same day, November 23, 1791. They were soon invaded by sealers, who almost exterminated the seal population. But since the island was made into a sanctuary for seals, their numbers have increased, and they are once again abundant. These islands see human visitors only rarely.

26. Auckland Islands

At 51° south, 166° east, the Auckland Islands are larger and more habitable than New Zealand's other sub-Antarctic islands. The plant and animal life here is comparatively lush for such a high latitude, and the islands have no permanent ice or snow. Lying about 200 miles southwest of Stewart Island, New Zealand, the group consists of one large and 5 smaller islands with several other islets and rocky pinnacles. The main island (Auckland), 179 square miles in area, is 24 miles long and from

3 to 16 miles wide, with Adams Island (35½ square miles) to the south, Enderby Island (1,770 acres) to the north, and Disappointment Island off the west coast. All these islands are of volcanic origin. The hills of Auckland rise to just over 2,000 feet. Its west coast is a long, unbroken line of high, steep cliffs. But the east coast is deeply indented with long, narrow inlets, which provide many good, sheltered anchorages, notably in Carnley Harbour, which divides Adams Island from the main island.

The climate is cool (35° to 65°), humid, cloudy and very windy. Nevertheless, most of the area is clothed with shrubby forest at lower elevations. Above about 300 feet there are open patches of tussock grass and sub-Antarctic meadows. Soils are peaty, waterlogged and sour.

For several decades after the group was discovered in 1806, whalers and sealers visited these islands regularly. Several scientific expeditions studied these islands between 1839 and 1842. In 1849, Charles Enderby sponsored an attempt at establishing a permanent whaling settlement on Auckland. But by 1852, the poor soil and marginal climate had defeated the settlers. In the 1890's, cattle and sheep were grazed here with some success, but the isolation of the islands led to the abandonment of this venture. A few wild cattle, pigs and goats still survive.

Oceanic birds such as petrels, penguins and albatrosses are plentiful here. There are also flocks of parakeets. The fur seals, which were hunted to near extinction, are now increasing in numbers. Sea lions breed here, and other sea mammals such as elephant seals, are regular visitors. As for the human population, there was a brief occupation during World War II, but since then these islands have remained uninhabited.

27. Bounty Islands

About 490 miles east of Stewart Island, New Zealand, at 48° south, 179° east, lie the Bounty Islands, a group of 15 small granite islets, with a total area of about one-half square mile. These are the barest, bleakest and most desolate of New Zealand's outlying islands. The terrain consists of bare rock, without any natural vegetation, and lacking any permanent fresh water supply. Vast numbers of seabirds, including penguins and mollyhawks, make their home here.

Captain William Bligh of the *Bounty* discovered and named these islands in 1788. Sealers began to work these islands in the early 1800's, and the vast seal population was soon almost completely destroyed. Today the seals are again increasing in numbers, but only very slowly. These islands are difficult to visit and so unattractive that no one is likely to stay there long.

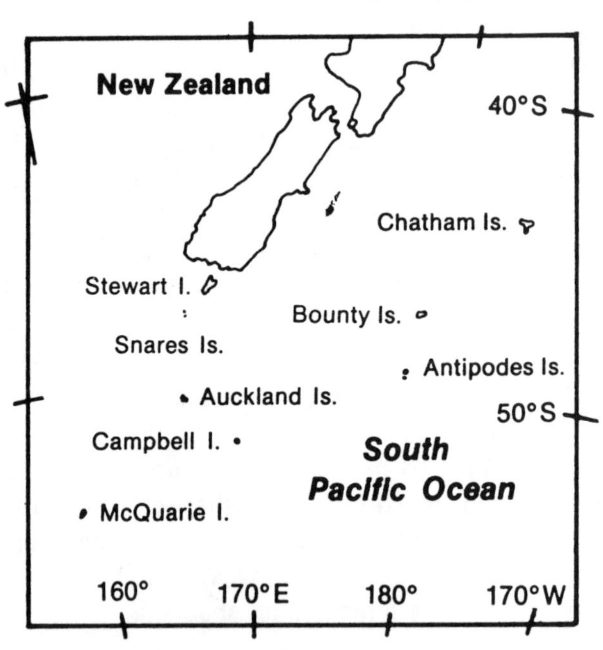

Sub-Antarctic Islands near New Zealand

28. Antipodes Islands

At 50° south, 179° east, which is almost the antipodes of London, England, lies this rocky, uninhabited group. The Antipodes have a total area of about 24 square miles and consist of one large island, Bollons, and several offshore islets and rocks. Bollons, 5 miles by 3 miles, is a plateau bounded by steep cliffs. Its highest point, Mount Galloway, rises to 1,320 feet. The surface is rough with a widespread blanket of waterlogged peat and swamps in the hollows. The dense cover of coarse tussock grass and hardy herbaceaous plants growing on the cold, wet peat indicates that climatic conditions are bleak and inhospitable.

These islands, which were once famous for their fur seals, were discovered in 1800 by Captain Waterhouse of HMS *Reliance.* The first sealer to work the islands was Pendleton, an American, whose gang was marooned there for nearly 2 years. They brought 60,000 skins back to Sydney, Australia, and started a rush of sealing on these islands that soon exterminated the stock.

A cache of supplies for the use of castaways was kept on Antipodes, but only 2 wrecks have occurred there. The crew of the first, *The Spirit of Dawn,* in 1893, didn't find the depot, but it was useful to the castaways from the *President Felix Faure* in 1908.

29. Campbell Island

Campbell is a solitary island 5 miles wide lying about 150 miles southeast of the Aucklands at 53° south, 169° east. It has an area of 44 square miles. High and rugged in the south, where it rises to 1,867 feet, it slopes off more gently to the north where smoothed ridges and open valleys suggest considerable recent glaciation. The east coast is broken by the 2 long, narrow, sheltered inlets of Perseverance and Northeast Harbours; the former almost cuts the island in two. Off the high cliff coasts of the west and south are several little rocky islets.

The climate of Campbell Island is similar to that of the Auckland group, though a little colder. Still, there is not much snow or frost. The west wind blows strongly, as it does on most islands at this latitude. Campbell has less cloud and more sunshine than the Aucklands and its soil is not as wet or sour. Over most of its area there is a cover of tussock with some scattered patches of dracophyllum. Herbaceous plants are much less varied and abundant than they used to be, largely due to the activity of sheep which were introduced at the beginning of the 20th century and now run wild. The seals and penguins usually found in these islands are now abundant on Campbell. In fact, it has one of the largest Rockhopper Penguin colonies in the world.

Frederick Hasselburg, captain of the sailing ship *製Perser-verance,* which was owned by the Sydney, Australia, firm of Campbell and Company, discovered the island in 1810. The seal population was soon reduced to the point where sealers lost interest in the island. It was visited by the Ross Expedition in 1840, and now and then by whalers up to the 1890's. In 1896, sheep were introduced and the island was more or less continuously occupied until the settlement was abandoned in 1931. Presently the Government of New Zealand maintains a weather station on Campbell as it has done since 1941.

30. McQuarie Island

Lying midway between Australia and Antarctica at 55° south, 159° east, McQuarie Island is a rocky mountain range rising from the sea. The main island is 21 miles long by 3 miles wide, with an area of 46 square miles. Its highest elevation is 1,421 feet. Offshore lie two small uninhabited groups known as Bishop And Clerk and Judge And Clerk. The surrounding waters abound in rock ledges and uncharted reefs on which many ships have foundered. Westerly winds blow almost the whole year and reach speeds over 100 miles per hour. Rain and fog occur on over 300 days a year. When the wind shifts to the south, it brings snow and freezing weather blowing in from Antarctica. But the island is generally free from ice and snow in summer. There are no trees, but plenty of scrubby vegetation, including tussock grass, and the rhubarb-like McQuarie Island cabbage. The sea near McQuarie is very rich in plankton, and giant kelp, a kind of seaweed, is also abundant. Other species that feed on these lower organisms, including seabirds, seals and penguins, are now plentiful since the Australian Government declared the island a wildlife sanctuary. Formerly, many species had been hunted to extinction on McQuarie.

This abundant wildlife has always been the main attraction of the island, first for commercial exploitation, and now for scientific study. It was the search for fur seals that led to the discovery of McQuarie in 1810 by Captain Frederick Hasselburg, commanding the *Perserverance.* In following years, sealers worked the island so thoroughly that the fur seals were completely wiped out on McQuarie. In the 1890's, oil works were set up on the island to boil down penguins to extract the oil they contain. Huge vats capable of holding 2,000 penguins at a time were used. The oil obtained was shipped to New Zealand where it was used for tanning and in the manufacture of soap. While this oil industry shut down long ago, many relics of these oil works can still be seen on the island today.

The famous Australian Antarctic explorer, Sir Douglas Mawson, conducted scientific research on McQuarie from 1911 to 1913, and set up the first weather station there. At his urging, the Australian Government banned sealing on the island in 1919. Since then, the depleted wildlife has vastly increased in number. Even the fur seal has returned to McQuarie. Up to 1,000 adult fur seals are sometimes seen there now, and a few fur seal pups have been born on the island in recent years. The west coast of McQuarie, which is battered by the fierce west wind, is one of the few places in the world where albatrosses breed. Other animals, which were brought to the island by sealers, have been increasing too, sometimes with harmful effects. Ordinary housecats let loose on the island have gotten wild and now live on penguin chicks and petrels. Rabbits, which were introduced by sealers as a food supply, have multiplied to the extent that they are endangering the ecological balance of the island. They crop the vegetation right down to the roots and kill it off, leaving a barren desert behind. To control them, scientists are now spreading diseased lice to kill off the rabbits.

Beginning in 1948, the Australian National Antarctic Research Expeditions have maintained several scientific research stations on McQuarie.

31. Peter I Oy

Peter I Oy lies 210 miles from Antarctica in the Bellingshausen Sea, at 69° south, 91° west. It is about 15½ miles long by 6 miles wide and is entirely covered with ice and snow. The island's highest peak is Lars Christensentopp, which rises to 4,003 feet. A Russian expedition under the command of Thaddeus von Bellingshausen discovered the island in 1821 and named it for the Czar Peter the Great of Russia. The first landing on the island was made in 1929 by a team of Norwegian scientists. Today, Peter I Oy is a Norwegian possession.

32. Scott Island

This tiny New Zealand island lies 310 miles from Antarctica at 67° south, 180° west, near the Ross Sea. It is only ¼ mile by ⅛ mile in size, is surrounded by cliffs, and is entirely covered with ice. Lieutenant Colbeck on the *Morning* discovered Scott Island in 1902 while on his way to relieve the *Discovery* expedition led by the ill-fated Captain R.F. Scott.

33. Balleny Islands

The 6 islands of the Balleny group, which include Sturge, Young, Buckle and Sabrina, lie about 160 miles off Antarctica at 69° south, 159° east. They are thickly covered with ice, and mountainous, with highest elevations ranging from 600 feet to over 4,000 feet. These islands were discovered in 1839 by John Balleny, a British sealer. Because they are almost inaccessible by sea, not much is known about them. During the 1963/64 season, a New Zealand expedition surveyed the Ballenys by helicopter and found them unsuitable for a scientific station.

Part III
Atlantic Ocean Islands

34. Deserted British Isles

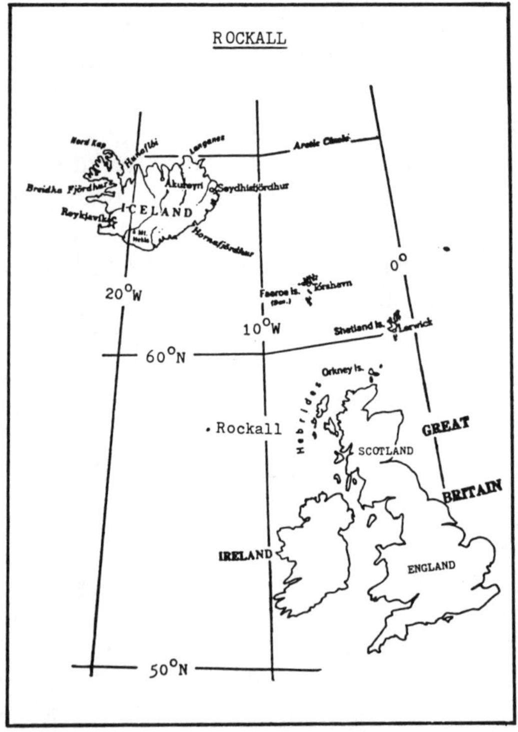

Rockall

a. *Rockall* is an uninhabited, 70-foot-high pinnacle of rock that lies in the North Atlantic south of Iceland and west of Scotland, 225 miles west of the Hebrides Islands, at 58° north, 14° west. Rockall was claimed by Britain as a territory in 1955. Such an isolated rock, inhabited only by seabirds, has no value in itself, but governments are eager to lay claim to such features because it gives them rights to natural resources in the ocean for 200 miles around.

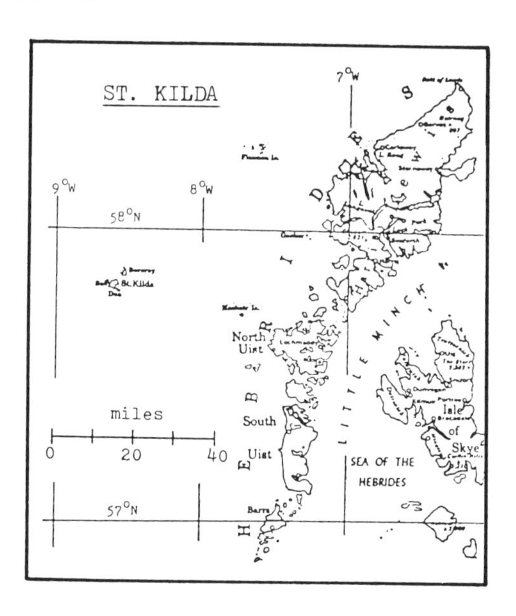

St. Kilda and The Hebrides

b. *St. Kilda* is the main island of the westernmost group of outer Hebrides, lying 42 miles northwest of North Uist, well away from the other Hebrides at 57° 57′ north, 8° 30′ west. St. Kilda is 3 miles by 2 miles, rising to 1,372 feet in the northeast, with precipitous cliffs except for a landing place on the south side. It was continuously inhabited from ancient times until 1930 when the surviving population of 36 was taken off, at their request, and moved to Scotland. It is now a bird sanctuary. The island, also known by its Gaelic name Hirta, has been the property of the Macleods clan for centuries. The group of islands also includes the smaller islands of Boreray, Dun and Soay. Boreray lies 4 miles northeast of St. Kilda and is one mile by one-half mile in size.

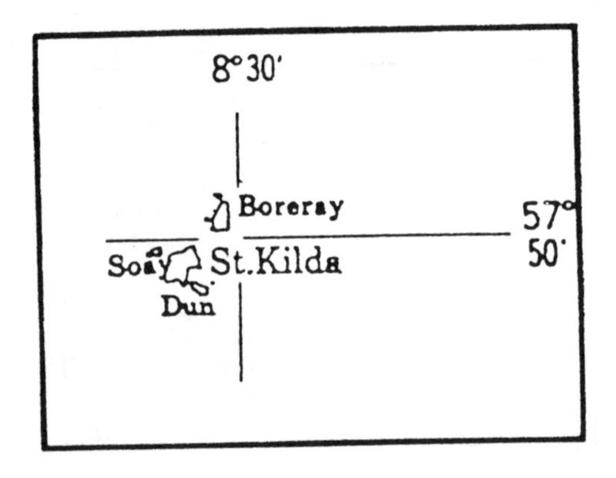

St. Kilda

Thousands of seabirds nest on these islands, so many that the cliffs of Dun and Boreray are white with bird excrement. No trees grow here so people built houses of stone with thatched roofs. The stone walls of the cottages still stand in the ruins of the village, but the roofs have all blown away. Without wood, they couldn't build boats, which left them stranded and isolated on these islands, and meant that they could only fish from the rocks. They also made extensive use of birds and birds' eggs, going so far as to use birds' beaks to pin together their thatched roofs. In historic times, sheep were introducted, improving the local economy for a while. Today sheep run wild, scrambling over the steep slopes like mountain goats. During the 1970's the

British Government maintained a rocket tracking station on St. Kilda.

35. Tristan Da Cunha

Tristan lies in the middle of the South Atlantic at 37° south, 15° west, with the neighboring uninhabited islands of Inaccessible, lying 12 miles to the southwest, and the 3 Nightingales Islands, 13 miles to the south-southwest. These islands are part of the British colony of St. Helena. Only Tristan is inhabited. The climate on these islands is warm and wet. Vegetation is lush, although the variety of species is limited. Seabirds are abundant along the shoreline. Tristan was discovered in the early 1500's by the Portuguese and the earliest anyone is known to have gone ashore was in 1643. Sealers hunted throughout the area in the 18th and 19th centuries. The first settlers came to Tristan in 1810 and there have been people living on the island most of the time since then. In 1961, volcanic eruptions forced the temporary evacuation of Tristan, but some Tristaners have since returned. Fishing for lobster and crawfish is the main industry.

36. Gough Island

Gough lies in the middle of the South Atlantic southwest of Cape Town, South Africa, and 230 miles southeast of Tristan Da Cunha, of which it is a dependency, at 40° south, 10° west. It has the same warm, wet climate as Tristan and is covered by thick vegetation. There are no large land animals, no amphibians, no reptiles, no land mammals, and no fresh water fish, but there are abundant seals, penguins, and seabirds. Gough has never been inhabited, and the native ecology has not been disturbed much by humans. There are large guano deposits that have never been worked. In 1955, a research team named the Gough Island Scientific Survey went out from Cambridge University, England, and spent 6 months on Gough investigating the wildlife and geography.

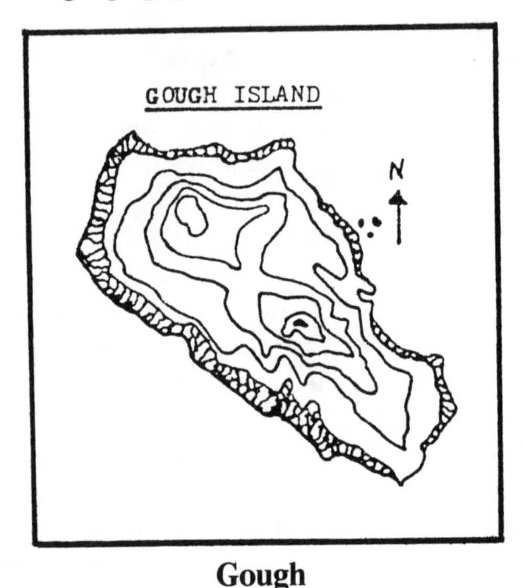

Gough

Gough has a considerable area, one of the larger uninhabited islands. It rises up to 2 peaks with a saddle in between, like a 2-humped camel. The higher southern peak is over 3,000 feet high. Steep cliffs surround most of the coast, but there is a natural harbor on the east side. Gough is a British possession and South Africa maintains an 8-person weather and research station there.

37. Falkland Islands

Some years ago it would have been safe to say that the Falklands were a little-known group of islands. But that was before world attention was focused on this bleak outpost of civilization by the war between Argentina and Great Britain. And these islands are a remote outpost indeed. The Falklands are one of the most southerly places on the earth permanently inhabited by real families, unlike the scientific stations in the Antarctic which have no permanent residents.

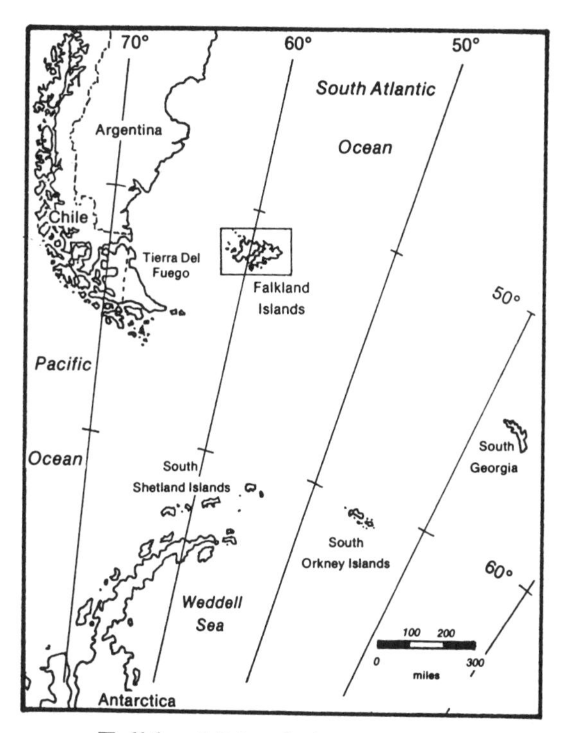

Falkland Islands in relation to South America and the Antarctic

The Falkland Islands, known as Islas Malvinas in Argentina, lie in the South Atlantic about 300 miles east of the southern end of Argentina at 51° to 53° south, 57° to 62° west. The group consists of 2 large islands, East Falkland and West Falkland, and about 340 smaller islands with a total area of

4,700 square miles. The islands have a population of about 1,800, about 1,000 of whom live in the town of Stanley, which is the capital. The Falkland Islanders are almost entirely of British descent.

These islands have a cool climate, with only a small variation in temperature throughout the year. The summer temperatures are similar to Scotland, but winter temperatures are warmer than Scotland, more like other parts of Great Britain. The Falklands' average temperature in summer is 49° F and in winter is 36° F. The highest temperature ever recorded was 79° F and the lowest was 12° F. Winds are strong — the average wind speed is 19 miles per hour — and they tend to blow from the west. The presence of South America upwind modifies the oceanic climate with the result that rainfall is reduced to only about 25 inches per year at Stanley. Frosts and snowfalls can occur in any month of the year, but they are rare in summer. Snow generally melts within a few days even during winter. There are only about 18 days a year when more than half of the sky is free of clouds. There are almost no trees, although small vegetation such as bushes and grasses are abundant. In gardens, root crops tend to grow best, and fruit is scarce except for certain berries.

Overland travel is difficult since there are no real roads outside Stanley. Most transport is by water around the coast. Much of the countryside is bog, except for areas of almost impassable sharp rocks called stone ruts. Another hazard found near the coast is elephant seal wallows — hard-to-see deep pits full of wastes, as treacherous as quicksand.

Only 15 of the islands that lie offshore of the 2 main landmasses are inhabited. Some of the uninhabited offshore islands are parts of privately owned farms, and sheep or other livestock are pastured on them. The rest of the uninhabited islands are valuable as breeding grounds for wildlife, especially penguins, seals and seabirds. Many of the offshore islands are protected as nature reserves or wildlife sanctuaries. All of the 14 islands in the Jason group, for example, are either reserves or sanctuaries.

There is no unclaimed land or islands in the Falklands. About 28,000 acres is Crown Land and the rest is owned freehold by various individuals and companies. About half the land is owned by the Falkland Islands Company, which is a subsidiary of the Coalite Group, Ltd., a publicly-held British corporation. The only important industry on the Falklands is sheep-raising, with wool as the major export item. The islands' sheep population runs about 600,000. About half of these animals are owned by the Falkland Islands Company. Also, there may by oil offshore, but oil hasn't actually been found yet. Oil companies have been hesitant to invest in test drilling because the Islands' future political status remains uncertain.

Since 1833 the Falklands have been a possession of Great Britain, and almost all of the present-day Falklanders consider themselves British. The Islanders are fiercely loyal to their mother country and almost unanimously want the Islands to remain a British Colony. But Argentina has never given up its claim to these Islands, which they call the Malvinas, based on an Argentinian colonization attempt in the early 19th century. This dispute over sovereignty has been a sore point between Britain and Argentina for years. But in recent years, before 1982, negotiations between Britain and Argentina had been proceeding and a more cooperative attitude seemed to be developing. The sticking point was that the Falklanders were adamant in their refusal to accept Argentine sovereignty, and Britain didn't want to make any disposition of the Islands without the consent of the inhabitants.

Suddenly, in April 1982, the situation changed dramatically. Using the salvage operation dispute on South Georgia as a pretext, Argentina invaded the Falklands on April 2 with about 5,000 troops, who quickly overwhelmed the 84 British Marines who were the only defense force on the Islands. The next day, Argentine forces also occupied South Georgia and captured the British Antarctic Survey personnel who were stationed there.

Making a strong response that surprised the Argentines, Britain immediately sent a large fleet steaming into the South Atlantic. The Argentines built up a force of about 10,000 troops on the Falklands. Fearing that the capital would be at the center of the coming clash, all except about 250 of the 1,000 residents of Stanley found refuge at inland sheep ranches for the duration of the war. Britain declared a sea and air blockade around the Falklands and, on April 25, launched a commando attack that recaptured South Georgia against stiff Argentine resistance.

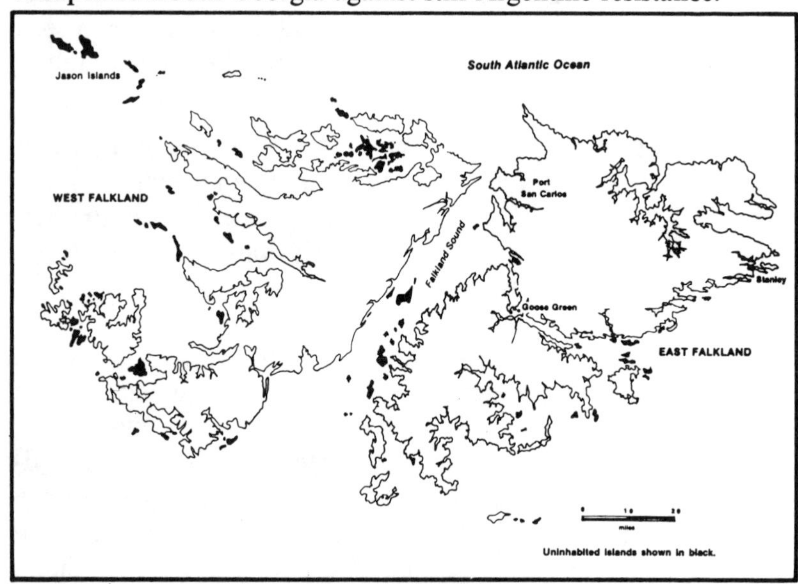

Falkland Islands

Turning to the Falklands, the British repeatedly bombed and shelled the airstrips of Stanley and Goose Green using naval guns and Harrier planes from its aircraft carriers, to cut the Argentine supply lines. On May 2, a British submarine torpedoed and sank the Argentine cruiser *General Belgrano,* second largest ship in the Argentine navy, as it was leading a task force toward the Falklands. Meanwhile, the Argentine air force continually attacked British ships, and on May 4, an Argentine fighter launched an Exocet missle that hit and destroyed the British destroyer *Sheffield.*

On May 21, British forces landed at Port San Carlos on the Falkland Sound side of East Falkland Island. The Argentine air force strongly opposed the landings and a fierce air battle raged over the Sound. Five British ships were damaged by Argentine aircraft, two seriously. But many Argentine planes were shot down and the landing was successful. By the next day, 5,000 British troops were ashore. Argentine ground troops gave a somewhat less stunning performance than their air force. The expected counter-attack against the British bridgehead never happened. And in subsequent battles, the Argentine troops, mostly poorly trained draftees, were easily overwhelmed by the professional British forces, and readily surrendered.

A few days after coming ashore, the British moved south and recaptured Darwin and Goose Green, the second large settlement in the Falklands, after Stanley, with a population of 82. The bulk of the Argentine forces fled back to Stanley, which they continued fortifying in preparation for the decisive battle. British troops moved on Stanley in two columns along northern and southern routes. Through early June they pushed the Argentines steadily back toward the capital. Finally, they drove the Argentine defenders off the high ground in front of Stanley and broke their defensive line. On June 14, the Argentines raised white flags and offered to surrender. The British recovered complete control of the Falklands.

In the 74 days of the Falkland Islands War, about 1,000 people lost their lives, including 234 British troops and 3 civilians (Falkland Islanders) killed during British shelling of Stanley. The financial cost was enormous. The South Atlantic fighting cost Britain $2.16 billion, according to certain London analysts, in the replacement costs of ships sunk or damaged and other material lost in the war. This is more than $1 million for every Falkland Islander. And the prospect of continued high defense costs stretches on into the indefinite future, since Britain must now keep a large garrison and fleet in the South Atlantic. At present, buoyed up by the euphoria of victory, the British Government says it will pay whatever it takes to defend the Falklands. But will this conviction remain as firm, once passing time has dimmed the thrill of victory, or after another government comes to power in Britain? Britain is now talking about the possibility of a multi-national force to defend the Falklands, with a role for the United States Government. This would mean that other nations would relieve Britain of some of the expense of defending the Falklands. Unless this is done, only one other long-term solution remains — Britain must come to terms with Argentina.

In the background, mostly unmentioned, there always remains the possibility of evacuating the Falklands. Britain could announce that, due to the high cost, after a certain date it would no longer defend the Islands. Financial assistance might be offered to any Falklander who wanted to relocate. Any reasonable amount spent on such a relocation would cost Britain less in the long run than continuing to keep a large military force in the South Atlantic. Perhaps arrangements could be made for the Falklanders to settle on one of the islands off Scotland, if they were willing, such as one of the Hebrides. The climate there is similar to what they are familiar with, and they could probably continue to find employment as sheep herders. This would put them in a place where Britain could easily defend them without incurring extraordinary expenses. Meanwhile, if the British Government were to walk away from the Falklands and leave undefended those Islanders who choose to remain, it can be taken for granted that Argentina will assume control in short order.

In the immediate aftermath of the War, talks between Britain and Argentina are out of the question. But after some passage of time, it seems inevitable that negotiations between Britain and Argentina over the future of the Falklands must resume and continue until some conclusion is reached. This much is clear: the Falkland Islanders can't expect British taxpayers to continue indefinitely to pay out such enormous sums for their defense, as were spent in this war.

38. South Georgia

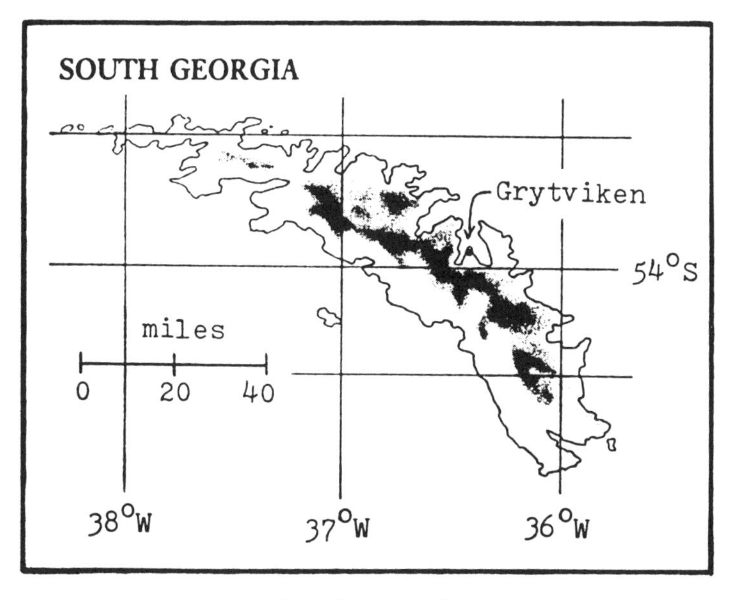

South Georgia

South Georgia is an island group in the South Atlantic consisting of one large island and a few small outlying islands. The main island, named San Pedro, but usually referred to as South Georgia, is mountainous and barren, about 120 miles long by 20 miles wide, with an area of 1,450 square miles. South Georgia is located at 54° south, 37° west, about 1,200 miles east of the southern tip of South America (Argentina) and about 800 miles southeast of the Falklands.

A chain of snow-covered mountains runs down the center of the island, with its highest point, Mount Paget, rising to 9,625 feet. Glaciers run down the valley and, in some cases, extend all the way to the sea. The climate is cold and damp with a mean annual temperature of 35° F, and a strong northerly prevailing wind. Vegetation is sparse. There are no trees and no native land animals, but sea birds and seals are abundant along the coasts, and reindeer have lived there since they were brought to the island in the early 20th century. Extensive kelp beds are found in the sea just offshore.

In the 1800's, sealers used South Georgia as a major base of operations. Later, several whaling stations operated on the island. From 1904 until 1966, when the last of the whaling stations, Grytviken, shut down operations, South Georgia was the most important whaling center in the Antarctic. Situated in a natural bay on the north side of the island, Grytviken once housed a population of up to 500, but is now inhabited only in summer by a small British Antarctic research team with a 6-man crew. There are great piles of abandoned machinery at every former whaling station, and entire ships lie rusting away in Grytviken harbor.

The first sighting of South Georgia was reported by a Spanish ship in 1756, but the Spanish didn't claim the island or name it. In 1775, while on his second voyage, Captain James Cook came upon the island and named it South Georgia in honor of King George III of Great Britain. Although South Georgia has been British since Cook's discovery, this didn't prevent Argentina from laying claim to it during the 1982 War, on the mere pretext that it is administered from the Falklands, over which they have a claim with a little more substance to it.

The 1982 War for the Falkland Islands between Britain and Argentina began with a dispute about a salvage operation on South Georgia. A few years ago, an Argentine scrap dealer named Constantino Davidoff heard about 3 whaling stations on South Georgia that had been abandoned in 1964. They were at Leith Harbour, Stromness and Husvik. In September, 1979, Davidoff signed an option to dismantle the 3 stations with the owner, a Scottish firm named Christian Salverson, Ltd. Davidoff obtained the proper clearance from the British and sailed to South Georgia for a few days in December, 1981, to look over the stations. A few months later Davidoff sent his crew to South Georgia to begin the salvage work. In preparation for that, he gave information to the British consul in Argentina on March 9, 1982, and apparently he thought all the legal and political formalities had been taken care of when his crew arrived at South Georgia on March 19. Upon arrival, the Argentine workers raised the Argentine flag over their camp. Why they raised the flag is not known. Perhaps it was a prank. British Antarctic research scientists at Grytviken, about 5 miles across the bay, saw the Argentine flag, called London on the short wave radio, and told the British Government, "The Argies have landed."

On March 22, the British Government protested to the Argentine Government that the scrap dealer had landed illegally. On March 24, the Royal Navy icebreaker *Endurance* was sent to South Georgia with a party of British Marines to force the Argentines to leave. Argentina then sent a navy ship of its own to "protect the interests" of the Argentine scrap metal workers. Britain sent another ship, the *John Biscoe,* with more marines. And Argentina sent an additional 5 warships.

Meanwhile, the rhetoric also escalated. Britain claimed that the Argentine workers should have stopped at the Falklands to have their papers stamped, since the South Georgias are a dependency of the Falklands. Argentina insisted that the workers had adequate documents, and in any case, they added, since the Falklands rightfully belong to Argentina, the South Georgias were Argentine also, so the salvage crew was actually on Argentine soil. Finally, Argentine military forces invaded the Falkland Islands on April 2, and they captured South Georgia on April 3.

Britain responded by sending a large naval task force into the South Atlantic. The first military action by this force occurred on April 25 when British forces disabled an Argentine submarine that was at Grytviken landing supplies for the Argentine troops. That same day, a British force landed at Grytviken and defeated Argentine forces, recapturing South Georgia, and reasserting British sovereignty over the island.

39. South Sandwich Islands

The nine South Sandwich Islands lie about 450 miles southeast of South Georgia at 56° to 59° south, 26° to 28° west, where they mark the eastern end of the Scotia Sea. The South Sandwich Islands consist of a southern group which includes the islands: Southern Thule, Bristol, Montagu, Saunders, Candlemas and Vindication, which were discovered by Captain Cook; and a northern group, known as the Traversay Islands, consisting of Leskov, Viokoi and Zavodovski, discovered by a Russian Expedition led by Bellingshausen. These islands, which have a total area of 120 square miles, are extremely barren, being mostly ice-covered, although there are numerous patches of bare rock. There is very little plant life and no land animals on these islands, but seabirds and penguins are numerous, and some seals have been seen. Due to sheer cliffs and heavy surf, landing by boat is difficult. The British used helicopters operating off ships to survey the islands in the early 1960's. A notable feature of these islands is their extensive volcanic activity. Around active fumaroles one finds oasies of plant life consisting of such low forms as mosses and liverworts. Perhaps it would be possible to drill geothermal wells to provide dependable controlled heat for an artificial environment, that might be built underground, which would be hospitable to human life.

Since these islands lie north of 60°, they fall outside the jurisdiction of the Antarctic Treaty. The South Sandwich Islands are British territory, administered from the Falklands. On June 19, 1982, in the last military action of the Falkland Islands War, British helicopters flew over a naval weather station that the Argentines had set up on Southern Thule, then they landed and surrounded the outpost. The ten men staffing the station were taken prisoner by the British forces. This short-lived Argentine outpost was the only habitation the South Sandwich Islands have ever seen.

40. South Orkney Islands

The South Orkneys consist of the large islands of Coronation and Laurie, and the smaller islands of Powell and Signy, plus a group of smaller scattered islands called the Inaccessible Islands, all located at about 61° south, 44° to 46° west, which is about 450 miles southwest of South Georgia. The islands have these dimensions: Signy is 5 by 3 miles; Powell is 7½ miles by 1¾ miles; and Coronation is 29 miles long. The whole group together covers an area of 240 square miles. These islands have a harsh, truly Antarctic climate, more like one would expect several degrees south of their actual latitude of 61° south, due to their location near the Weddell Sea, which is always covered with ice. Of these islands, only Signy, which lies south of the much larger Coronation Island, is relatively ice-free. There is a British research station on the east side of Signy and an Argentine station (Orcadas) on Laurie. The South Orkneys lie in the region of the Antarctic that is claimed by both Argentina and Britain, but these rival claims are suspended for the duration of the Antarctic Treaty.

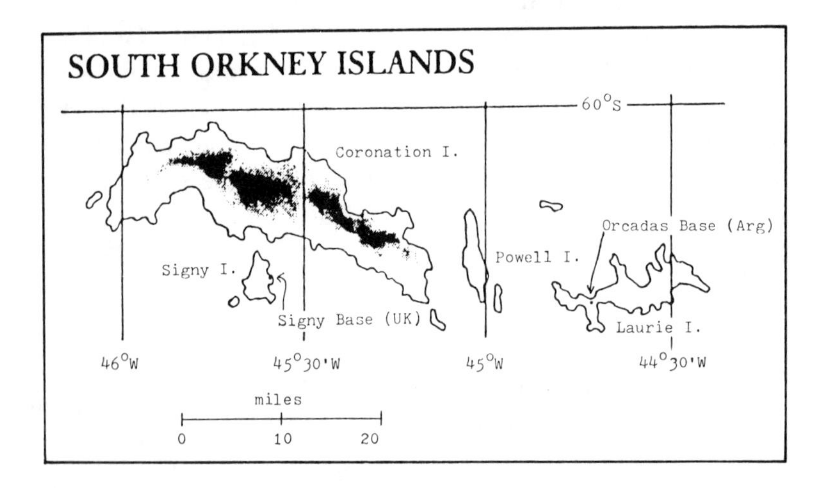

South Orkney Islands

41. South Shetland Islands

The South Shetlands are found at 61° to 64° south, 54° to 63° west, lying off the coast of the Antarctic Peninsula. These islands are mostly mountainous and ice-covered. The most unusual member of the group is Deception, which is round, about 9 miles in diameter, rises to 1,000 feet high, and is volcanically active. The center of the island consists of an open crater that forms a sheltered harbor, which was formerly a popular gathering place for sealers. Deception is relatively ice-free, and like all of the South Shetlands, it has a mild climate for the Antarctic, with considerable plant life, mostly mosses and lichens. Hot sulphur springs are found on Deception's beaches. Argentina, Chile and Britain all had stations on Deception, but they were quickly evacuated in 1967 when a volcanic eruption occurred. Now Argentina alone maintains a summer base on Deception.

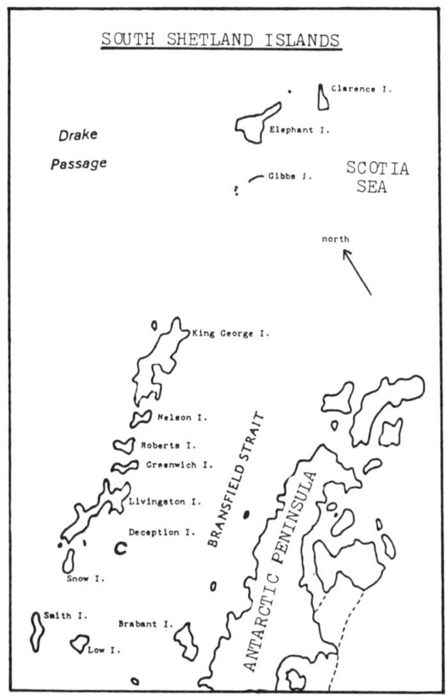

South Shetland Islands

King George Island, the largest of the South Shetlands, has more research stations than any other island. At last count there were 7 stations operating there: Bellingshausen (USSR), Arctowski (Poland), Artigas (Uruguay), Teniente Rodolfo Marsh (Chile), Great Wall (China), Teniente Ulbany (Argentina), and Commandante Ferraz (Brazil). There was also formerly a British station at Admiralty Bay, which is now closed. These scientific stations are on other South Shetland Islands: Astronomo Cruls (Brazil, summer only) on Nelson Island, Capitan Arturo Prat (Chile) on Greenwich Island, and Engenheiro Wiltgen (Brazil, summer only) on Elephant Island.

The South Shetlands lie in the region where overlapping disputed territorial claims have been made by Britain, Argentina and Chile, but these claims are in suspension presently under the terms of the Antarctic Treaty.

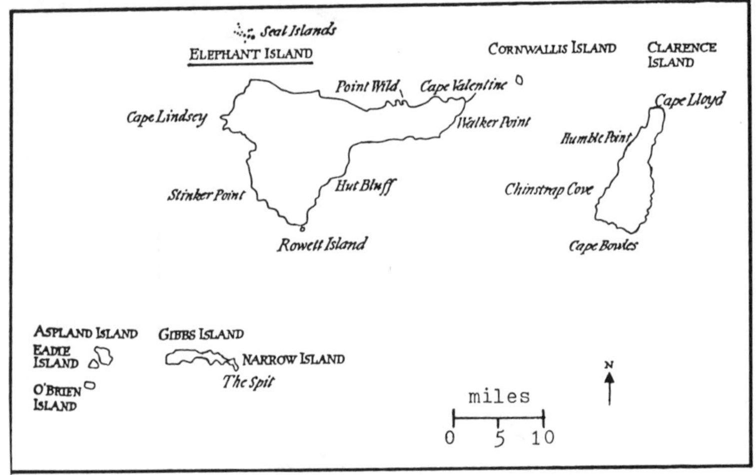

Elephant Island Group

Elephant Island, in the South Shetland group, played a part in one of the most remarkable episodes of Antarctic heroism and survival. On December 5, 1914, a British expedition led by Sir Earnest Shackleton sailed into Weddel Sea from South Georgia Island aboard the *Endurance.* Their objective was to cross the unexplored Antarctic interior to Ross Island. But the ship never reached the coast. In January, 1915, the sea froze around the ship, and the *Endurance* begin drifting westward with the ice.

For several months the ship remained frozen into an ice floe about 3 miles square and it was in no great danger. But by July, the wind-driven ice floe began to crack and buckle into pressure ridges around the ship. Conditions of the sea ice grew worse over the next couple months. The ice floe cracked through at the ship, and *Endurance* was lifted and twisted and squeezed by the immensely powerful ice. By October the ship was taking on water as pressure on the hull increased. On October 16, boats, sledges and provisions were moved out onto the ice floe. The next day *Endurance* was abandoned. Then the ice floes crushed and destroyed the ship. Nine months after being seized by the ice, the party of 28, still safe, was adrift on the pack of ice with a few sledges, 3 lifeboats, and limited supplies.

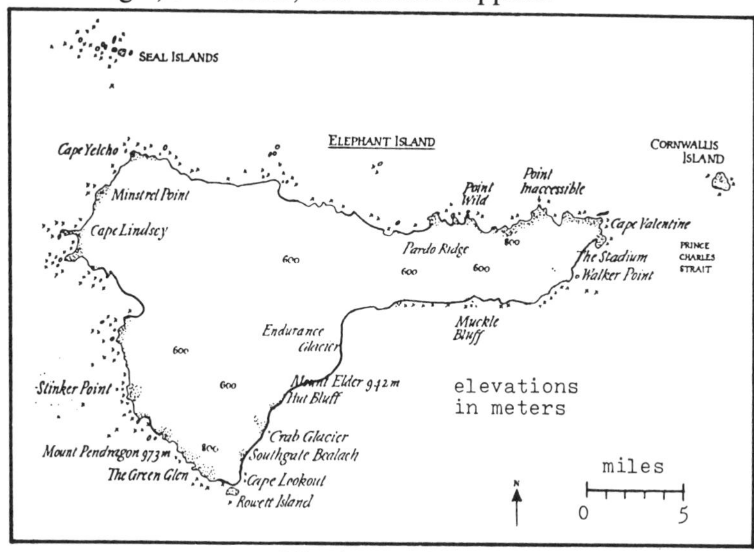

Elephant Island

The nearest food and shelter was at Paulet Island, 346 miles across the drifting ice from the stranded *Endurance* party. After packing food and camping gear and putting the boats on sledges, they set out across the ice. It was nearly impossible to drag the heavy boats through the soft snow, to chop passes through the pressure ridges, and to negotiate cracks between floes. The dogs were shot because of a food shortage. Strict rationing as the days moved into weeks and months was broken only when an occasional penguin or seal was killed to supplement their diet. They conserved energy by camping on the sea ice at what was aptly called *Patience Camp.*

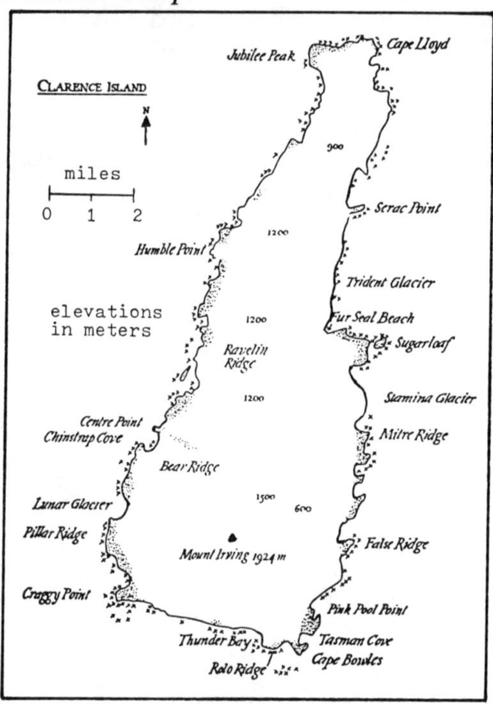

Clarence

On March 23, 1916, after 5 months of struggling to cross the driftng sea ice, Shackleton sighted Joinville Island, near Paulet Island, the ultimate destination for help, food and shelter. Three days later the pack ice opened, with leads of open water running in all directions. The boats were launched, everyone piled aboard, and the party began threading its way toward land.

On April 15, 1916, 15 months after besetment, Shackleton's group reached Elephant Island, just north of the Antarctic Peninsula. The ice leads, wind and current had carried them about 130 miles northwest of their intended goal of Paulet Island.

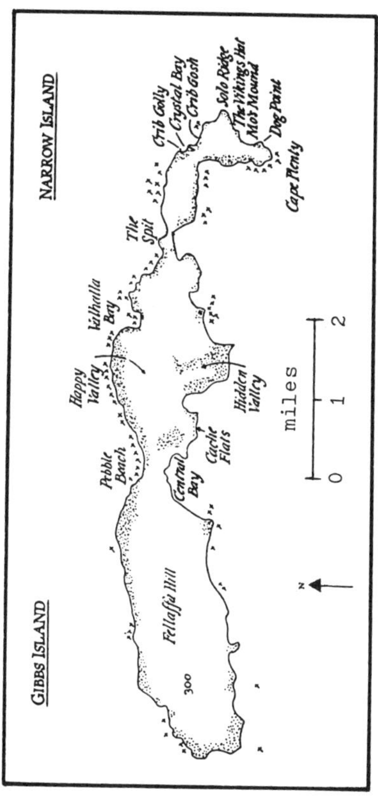

Gibbs

Nine days later, Shackleton and 5 companions began one of the most amazing small boat voyages in history, knowing that the lives of 22 persons left behind on the rocky shores of Elephant Island hinged on their success. The destination was a whaling station on South Georgia Island, 870 miles to the northeast. The 22½ foot lifeboat had to cross one of the stormiest seas in the world. For 2 weeks they struggled through gales before sighting the black cliffs and icy summits of South Georgia Island on May 8. The next day a landing was made on a rocky beach, but they were separated from help by the island's high central mountain range.

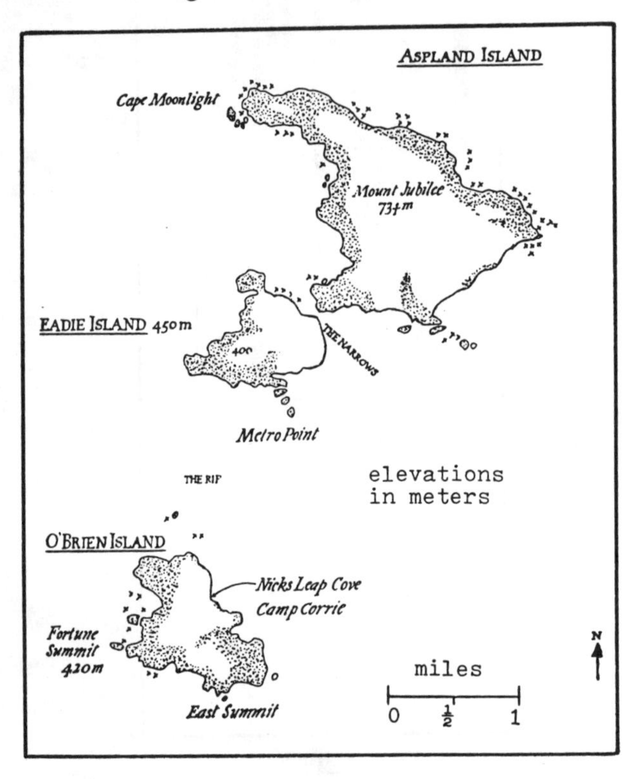

Aspland, Eadie, and O'Brien Islands

Shackleton and 2 companions set out over the mountains to reach the whaling station about 20 miles away. They sighted the whaling station after 2 days of climbing and descending perilous ice falls. The first people they met backed away in alarm, until they learned that the shaggy and emaciated figures before them were Sir Earnest Shackleton and members of his expedition.

A whaling ship was sent immediately to fetch the 3 others on the opposite side of South Georgia. The rescue of the main party on Elephant Island was not as simple. Five relief expeditions were launched before the Chilean steamer *Yelcho* got through and succeeded in the rescue. The men had been on Elephant Island for almost 5 months when their rescuers arrived.

Shackleton's success in leading his expedition to safety without losing a single person, despite great hazards, was perhaps a greater achievement than if he had successfully crossed the Antarctic continent.

After the departure of Shackleton's party, hardly anyone visited Elephant Island for over half a century, until December 1976 when a British scientific survey team conducted research in the Elephant Island group. They vistited all the islands in the group, studying wildlife and mapping the islands accurately for the first time, and afterward, Chris Furse, the team's leader, wrote a book, *Elephant Island, an Antarctic Expedition,* about their experiences.

In January 1985, Brazil built a summer-only research station on Elephant Island, which is the first research station to be set up on any of the islands of the Elephant Island group.

42. Bouvetoya

Located at 54° south, 3° east, Bouvetoya is the most southerly island on the Mid Atlantic Ridge. It consists of a single cone of volcanic origin that rises to 3,068 feet and covers 19 square miles. Most of the island is covered by an ice cap, which runs all the way down to the sea in places, and is usually hidden from sight by a thick cloud. Snow falls frequently. Temperatures average about 29° F in winter, and seldom rise above 35° in summer. Vegetation is limited to mosses and lichens. The island is inhabited mainly by seals and penguins, but no people.

The French explorer Bouvet de Lozier discovered Kapp Circoncision, which is the northwest point of Bouvetoya, in 1739, but he thought it was part of an unknown continent. Many expeditions passed by the island over the years, but there was no landing until 1927 when scientists from Norway went ashore to study the island. Later, Norway annexed Bouvetoya. South Africa has tried, but so far has not succeeded, to establish a weather station on Bouvetoya.

Part IV
Indian Ocean Islands

43. French Islands Near Madagascar

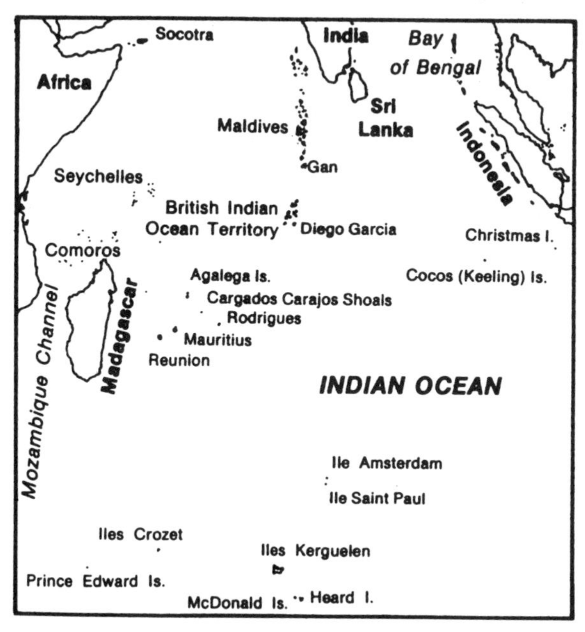

Indian Ocean

The Prefect of Reunion, who is the appointed representative of the French Ministry of Overseas Departments and Territories, serves as the chief executive of the heavily populated island of Reunion. He also serves as the administrator for several French overseas territories in the Indian Ocean — the islands of Ile Tromelin, Iles Glorieuses, Ile Juan de Nova, Ile Europa and Bassas da India. Three of these islets, Ile Juan de Nova, Ile Tromelin and Ile Europa, are now wildlife sanctuaries. Recently, the President of Madagascar has restated his country's claim to the islands in the Mozambique Channel — Juan de Nova, Europa and Bassas da India.

a. *Iles Glorieuses.* Iles Glorieuses consists of two main islets which have a combined area of about 1,090 acres. The larger islet is about 1½ miles in diameter and the smaller is less than ½ mile across. They lie 6 miles apart, separated by shallow sand banks that are above sea level at low tide. Iles Glorieuses is at 12° south, 47° east, northwest of the northern tip of Madagascar. The main island is covered with lush vegetation and coconut palms. It has an airstrip and a weather station. The weather station crew are the only inhabitants. The lagoon is used by snorklers.

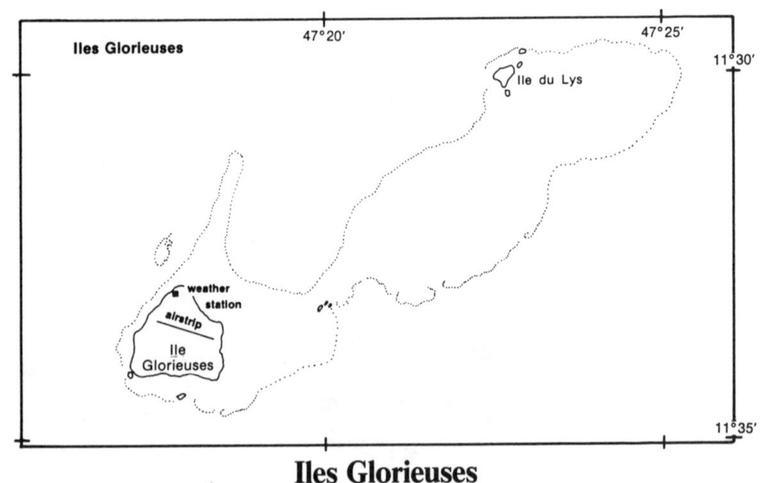

Iles Glorieuses

b. *Ile Tromelin.* Ile Tromelin is a single, oblong islet at 16° south, 54° east, east of northern Madagascar. It is about one mile long by ½ mile wide, with an area of about 200 acres. Fringing reefs surround the islet, which is sandy with scattered bushes. It has an airstrip and a weather station, whose crews are its only inhabitants.

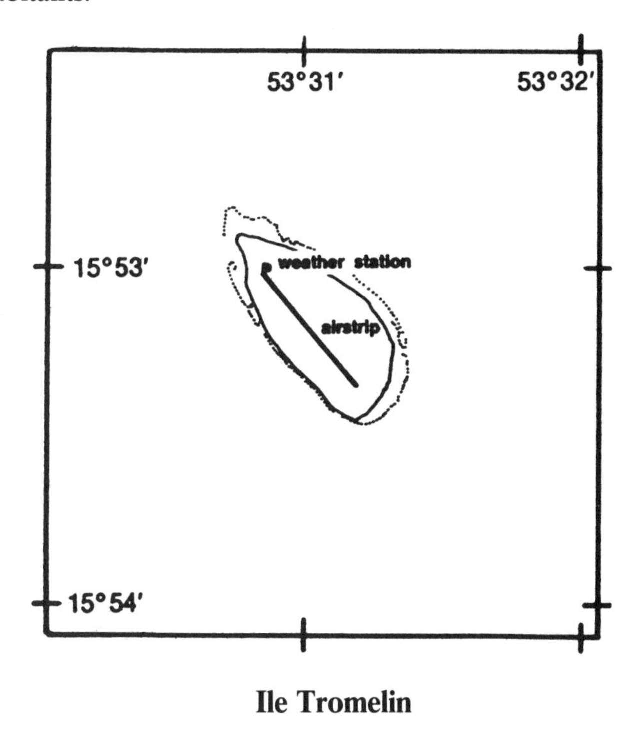

Ile Tromelin

c. *Ile Juan de Nova.* Ile Juan de Nova is a long narrow islet, about 3½ miles long by ½ mile at its widest point, giving it an area of about 1,150 acres. The islet lies in the Mozambique Channel at 17° south, 43° east. An extensive reef runs out to 2 miles and more offshore. Trees cover nearly the entire islet and there is a small railroad running down to a jetty which was built during the time when there were guano works on the islet. It also has an airstrip and a weather station, but no population except for the weather station personnel.

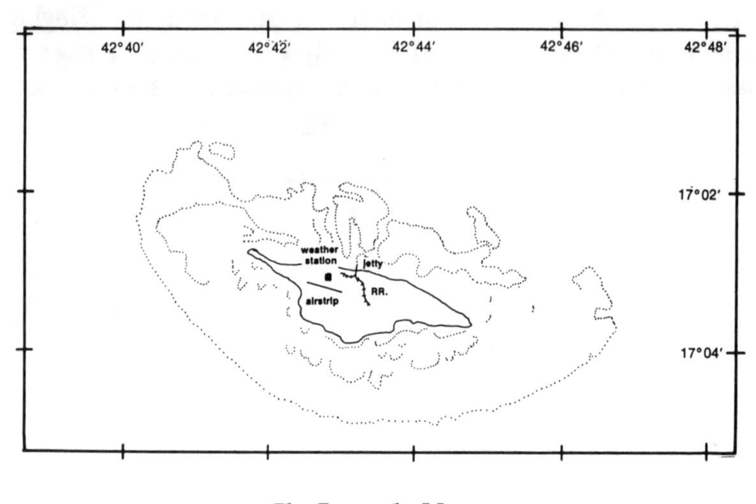

Ile Juan de Nova

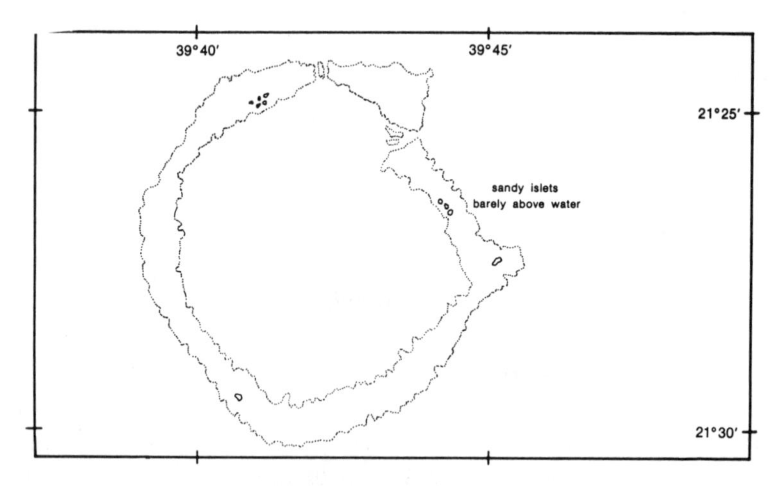

Bassas da India

d. ***Bassas da India.*** Bassas da India is a bare reef about 7 miles in diameter. There are a few tiny sandbanks barely above water but there is really no substantial land area. The reef has no inhabitants at all. It lies at 21° south, 40° east, in the Mozambique Channel, about 70 miles northwest of Ile Europa.

e. ***Ile Europa.*** Ile Europa is a single islet with a large bay opening toward the north, surrounded by a reef. It is about 4 miles in diameter and it has an area of 7.8 square miles. The islet is situated at 22° south, 40° east in the Mozambique Channel. It is covered with trees and bushes, and it has a weather station and an airstrip. The weather station crew are the only inhabitants.

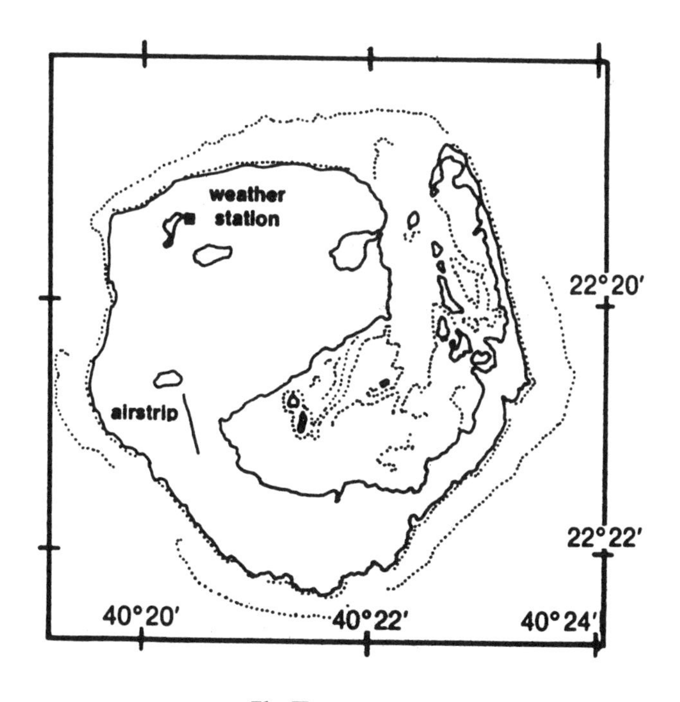

Ile Europa

44. British Indian Ocean Territory

The British Indian Ocean Territory (BIOT) was created in 1965 to provide sites for joint UK/US military facilities. The British transferred 4 of the least populated island groups from 2 of its crown colonies — the Chagos Archipelago from Mauritius, and the Aldabra, Desroches and Farquhar groups from the Seychelles — to form the territory. The latter 3 were returned to the Seychelles when it became independent in 1976. The establishement of the BIOT was prompted by a concern for the security of communications and transit routes between Africa and the Far East. Military facilities in the BIOT have since been opposed by nearly all of the island and littoral countries of the Indian Ocean, most of which support the UN resolution favoring a "zone of peace" in the Indian Ocean.

The sole remnant of the original BIOT, the Chagos Archipelago, runs from about 5° to 7° south, at 72° east, just below the equator, near the center of the Indian Ocean, and about 1,000 miles south of the southern tip of India. The Chagos islands are uninhabited, since their former population of less than 500 contract copra workers and their families were resettled on Mauritius and the Seychelles by the British. Diego Garcia, the largest of the Chagos islands, is the site of a recently completed US communications center, jointly run with the British, which serves mainly as a relay station for military traffic. The US is planning to expand the station to include limited air and naval support facilities, which involves extending the present 8,000 foot runway to 12,000 feet, dredging the lagoon, and increasing the fuel storage capacity. The enlarged installation will be used to supply navy ships crossing the Indian Ocean and to provide a refueling base for jet aircraft.

The island groups that make up the Chagos Archipelago consist of, from north to south: the atoll Peros Banhos; Salomon

Islands; the isolated Nelsons Island; Three Brothers; Eagle Islands; Danger Island; Egmont Islands; and the atoll Diego Garcia. The closest of these uninhabited islands to Diego Garcia with its militry facilities is Egmont Islands, which lie 77 miles from Diego Garcia, and the farthest away is Peros Banhos, 130 miles distant.

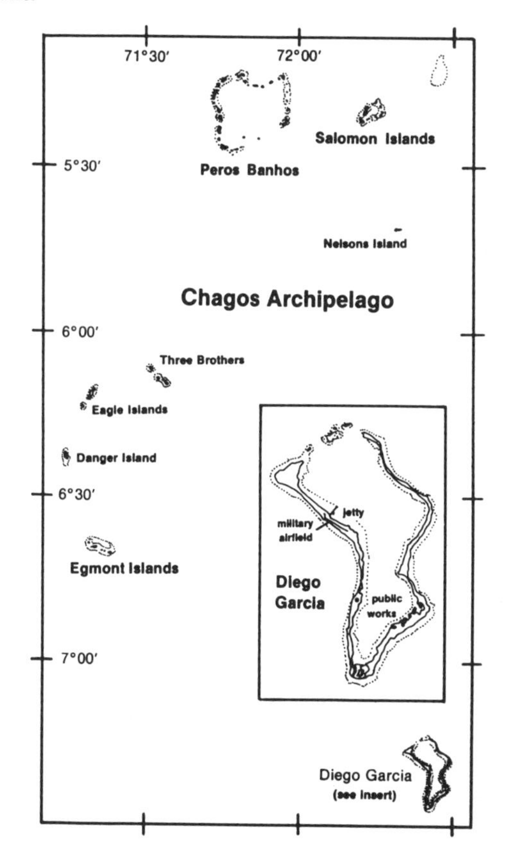

Chagos Archipelago (with Diego Garcia Inset)

Peros Banhos is a coral atoll at 5° south, 72° east. It has an area of about 4 square miles, and its reef encloses a lagoon 15 miles long by 11 miles wide. Its coconut plantations are now abandoned.

The Salomon Islands make up a coral atoll about 2 square miles in area at 5° south, 72° east. It has a lagoon about 5 miles long by 3 miles wide. There was formerly a settlement on the islet Ile Boddam and coconut plantations extend throughout the atoll.

The Three Brothers Islands, at 6° south, 72° east have a total area of 345 acres.

Egmont Islands, also known as Six Islands, is a coral atoll located at 7° south, 71° east. It has a lagoon 5 miles long by 1½ miles wide, and a total area of 644 acres, much of which is planted in coconuts.

The Chagos Islands are out of the path of cyclonic storms common to latitudes farther south. They are the wettest coral islands in the Indian Ocean, with up to 145 iches of rainfall annually, but the almost constant winds help make the islands' high temperatures and humidity more bearable.

45. Cocos Keeling Islands

On the eastern side of the Indian Ocean, 1,750 miles northwest of Australia, at 12° south, 97° east, lie the Cocos Keeling Islands, which consist of 2 coral atolls that are a territory of Australia. All of the 435 inhabitants (1978), as well as the economic resources and communications facilities, are on the southern atoll, the South Keeling Islands. North Keeling, a separate atoll that lies 15 miles to the north, is habitable, but uninhabited. North Keeling consists of a low ring of land, about one and one-third miles long by ¾ mile wide, surrounding a shallow lagoon. Total area of the atoll is about one square mile. There is no navigable passage through the reef into the lagoon, and the atoll has no good anchorage.

The Cocos Keeling Islands have been run for almost 150 years as a private estate of the John Clunies-Ross family, who are sometimes called the "White Kings of Cocos." The Clunies-Ross estate, situated on Home Island in the South Keeling atoll, still manages the coconut groves located on South, Home, West and Horsburgh Islands (all in the South Keeling atoll). However, the political sovereignty of the family ended in 1955 when the islands became an Australian possession. The estate workers are quartered on Home Island. The official representative of the Australian Government and most of the 140 members of the Australian community reside on West Island, the site of the airfield formerly used by the Qantas and South African Airways, but now maintained only for emergency and other unscheduled use. The two communities have little contact with each other.

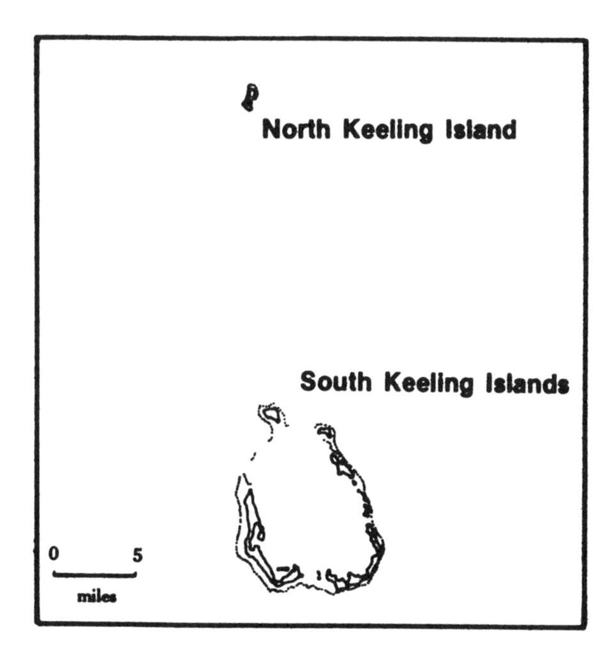

North and South Keeling Islands

The islands are low, averaging 3 to 16 feet above sea level, and the thin, poorly developed soil supports a varied cover of tropical vegetation. Shallow wells, the only source of fresh water, provide sufficient potable water for the present needs of the islanders. The tropical climate is marked by relatively constant temperatures, averaging 81° F, and high humidity throughout the year and by only slight seasonal variations in rainfall and wind direction. The southeast trades that blow for about 9 months of the year tend to ameliorate the discomfort of the high temperatures. Destructive storms are rare.

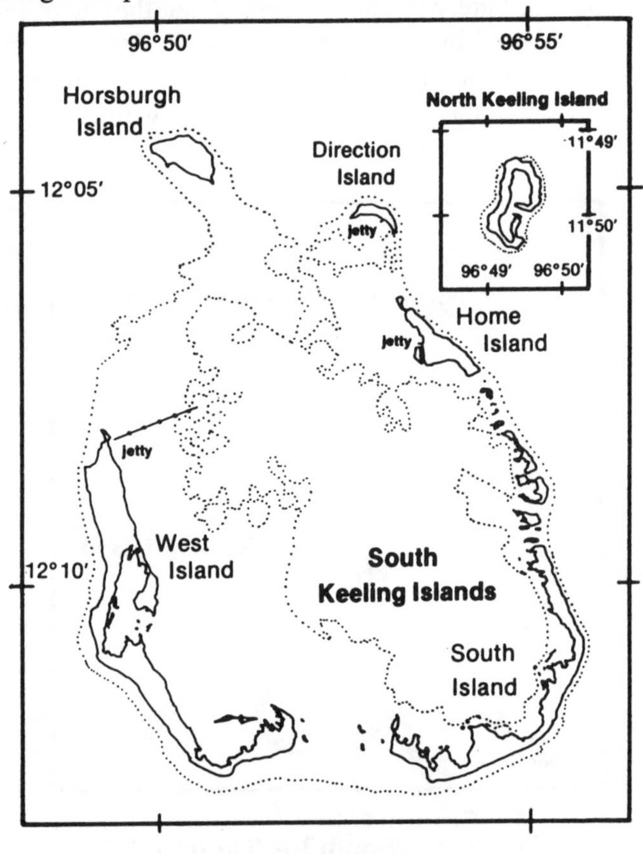

South Keeling Islands
(with North Keeling inset)

Copra (dried coconut) production is the basis for the economy of the islands; some 300 tons are processed each year and shipped to Australia. The work force consists of "Cocos Malays," Malay-speaking Moslems whose forebears were imported from the Netherlands East Indies and Malaya during the period 1827 to 1831. By the end of World War II, the Cocos Malay population was larger than the Clunies-Ross estate could support. A program of controlled emigration at that time resettled 1,600 Cocos Islanders on Christmas Island and in Sabah, now part of Malaysia. Since 1951, population growth has been restrained by a vigorous birth control program and by emigration when the worker population exceeds 500, the number considered optimum for the present community.

The administration of the Clunies-Ross estate has been the subject of criticism, by both the Government of Australia and a United Nation Special Committee, centering on the tight control the estate maintains on the lives of its workers. In 1975, the United Nations Special Committee urged Australia to seek ways to diversify the Cocos economy and to establish an identity for the Home Island workers as a community separate from the Clunies-Ross estate. Provisions have now been made to establish a judicial system, to upgrade the education of the islanders, and to elect a local government authority. As an added measure, the Government of Australia is negotiating for the purchase of the Clunies-Ross estate.

46. South Indian Ocean Islands

The islands of the southern Indian Ocean are located almost as far from civilization as man can travel on this planet. There are some scientific research stations on these remote islands but no settlements — no "natives." There are no airfields, and only an occasional supply ship calls. Situated in a belt of nearly perpetual storms and high seas, these desolate, bleak and windswept islands provide a habitat more suited to seals and marine birdlife than to man.

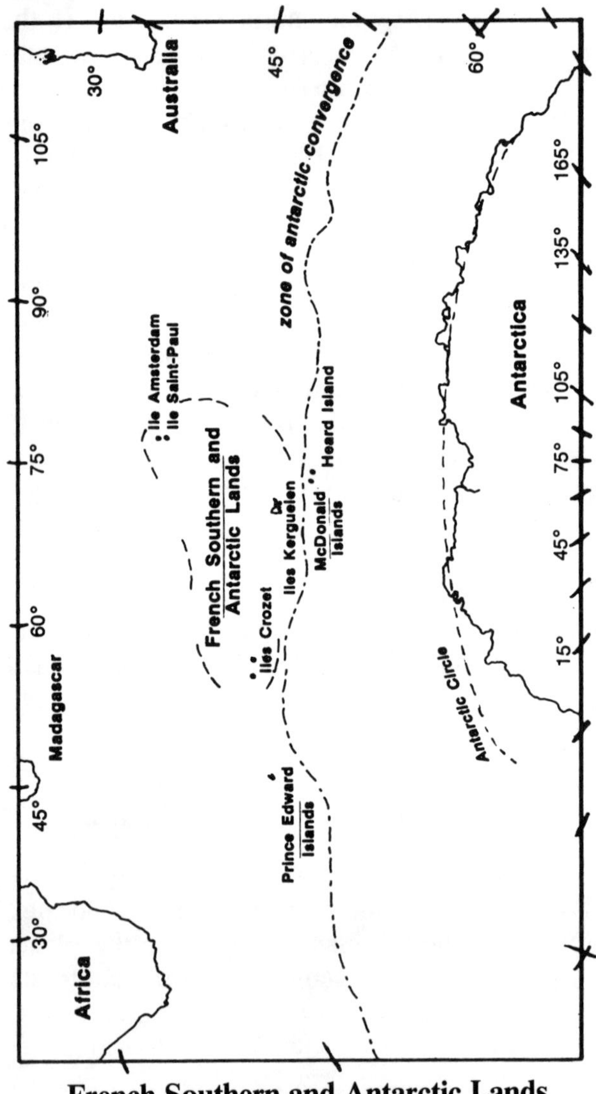

French Southern and Antarctic Lands

The islands are now of value principally as sites for scientific research. Marine studies are facilitated by the location of most of the islands in the zone of Antarctic convergence, where the cold waters of the Antarctic seas underride the warmer waters of the main body of the Indian Ocean. The islands also provide sites for the study of upper atmospheric physics and for the collection of meterological data in the seemingly limitless ocean expanse. Their remoteness makes possible the study of ecosystems that have been relatively undisturbed by man.

All of these islands are of volcanic origin and, although widely separated, have similar terrain, flora, fauna and climate. Low cliffs and rocky shores make landings from the sea difficult if not hazardous. Only the fiord-like inlets of the Kerguelens offer protected anchorages. Most cargo and personnel are transferred between ship and shore via barge or helicopter.

Most of the research stations are on the leeward sides of the islands, where there is some shelter from the prevailing westerly winds. The incessant battering by the elements deteriorates buildings rapidly. Reseachers are usually rotated after tours of 12 to 18 months to avoid mental depression that may be induced by the confined and lonely existence. Various other hazards to field research have been noted. It is easy to become disoriented and lost since compass readings are unreliable — the needle often rotating erratically within short distances — and low visibility conditions develop quickly and frequently. Discomfort associated with the windchill factor is ever present, often compounded by exposure to wind-driven sleet or snow and by unexpected soakings from loss of footing in boggy areas.

Commercial activity is limited to fishing near the Kerguelens and lobster trapping off Amsterdam and Saint Paul. The islands attained some economic importance during the 19th century in the heyday of fur sealing and whaling, but uncontrolled slaughter greatly reduced the seal herds, and advances in technology obviated the need for whaling ships to call at the islands.

Most of them have been made wildlife sanctuaries, and the threatened species are now re-establishing themselves.

Administratively, these islands are controlled by three countries. The French Southern and Antarctic Lands (Territoire Australe et Antarctique Francais — TAAF) comprises the largest and most important islands — the Kerguelens, the Crozets, Amsterdam and Saint Paul. The Prince Edward Islands belong to South Africa, and Heard Island and the McDonald Islands are Australian territories.

47. Prince Edward Islands

In the southern Indian Ocean lies the Prince Edward Islands group, at about 47° south, 38° east, 1,200 miles southeast of Capetown, South Africa. This group, a possession of the Republic of South Africa since 1947, consists of two islands which are the double peaks of an underwater volcano. The larger island is Marion, about 112 square miles in area, which rises to an altitude of 4,034 feet. The smaller of the two, with an area of about 17 square miles, is Prince Edward Island, which lies about 14 miles north of Marion. The highest elevation on Prince Edward Island is 2,204 feet.

Prior to 1965, little was known about these islands. In that year, a South African expedition visited the islands as part of a broad Antarctic research program. The group stayed several months, mostly on Marion Island. A permanent weather station has since been established on Marion, with quarters to accomodate as many as 30 researchers. There are unconfirmed reports that South Africa has recently built a missile facility on Marion Island.

Some 130 low volcanic cones are scattered over the central highland area of Marion Island. They alternate with shallow, glacier-formed basins and depressions in which ponds and extensive bogs have formed. Natural drainage is further retarded in the narrow coastal lowlands by wallows made by molting seals. The terrain on Prince Edward Island is similar, making foot travel on both islands precarious.

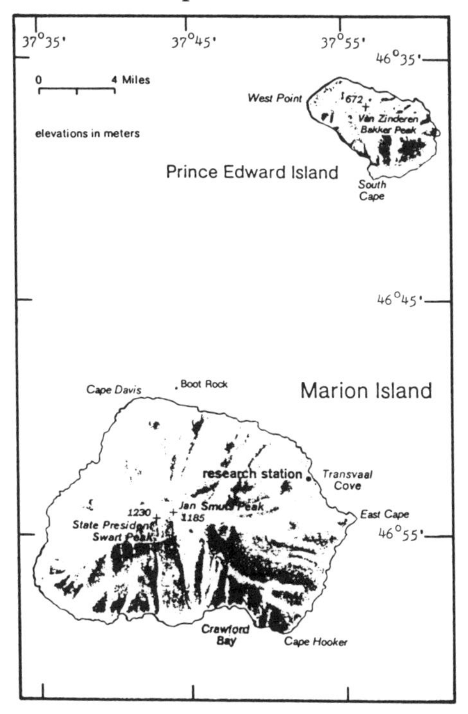

Prince Edward Islands

The prevalent weather condition on these islands is unpleasantly cool and wet. All too brief periods of sunshine are not sufficient to dispel the sense of continual cloudy weather. Fog or some form of precipitation, normally rain, but occassional drizzle or snow, occurs on an average of 311 days a year, and the relative humidity is high. The temperature permits seasonal herbaceous growth but is not warm enough for the establishment of trees or other woody plants. The vegetation consists mostly of such things as grasses and mosses and Kerguelen cabbage. Accompanying the damp, chilly atmosphere and gray skies are strong prevailing northwest winds that average 20 miles per hour and frequently reach gale velocity.

Much of the current research is concentrated on the ecology of the birds that breed on the islands — albatrosses, terns, petrels, prions and penguins. Large rookeries of both albatrosses and penguins abound along the rocky coasts. Farther inland, petrels and other species nest in pockets burrowed in the soft gray and black lava soils.

48. Crozet Islands

The Crozet Islands are a French possession, part of the French Southern and Antarctic Lands, that lie about 1,500 miles southeast of Africa at 47° south, 38° east. The archipelago consists of two groups of islands separated by about 60 miles of open sea. The eastern group, Ile de la Possession and Ile de l'Est (East), have permanent snow cover on their fog-shrouded higher peaks, while deeply eroded lower slopes bear a spare cover of tussock grasses, lichens and mosses. The western group consists of one large island, Ile aux Cochons (Island of Hogs), and two groups of smaller islands named Iles des Apostres (Apostles) and Iles des Pingouins (Penguins), that are little more than rocks and reefs serving as rookeries for a large population of seabirds. Ile aux Cochons is similar in terrain and vegetative cover to Ile de la Possession and Ile de l'Est. These islands are mountainous and have these highest elevations: Ile aux Cochons 2,526 feet, Ile de la Possession 3,064 feet, and Ile de l'Est 6,500 feet. The total area of the archipelago is 132 square miles. The climate is marked by constant westerly winds since these islands are in the "roaring forties" of latitude. The coasts are crowded with seals, penguins and seabirds.

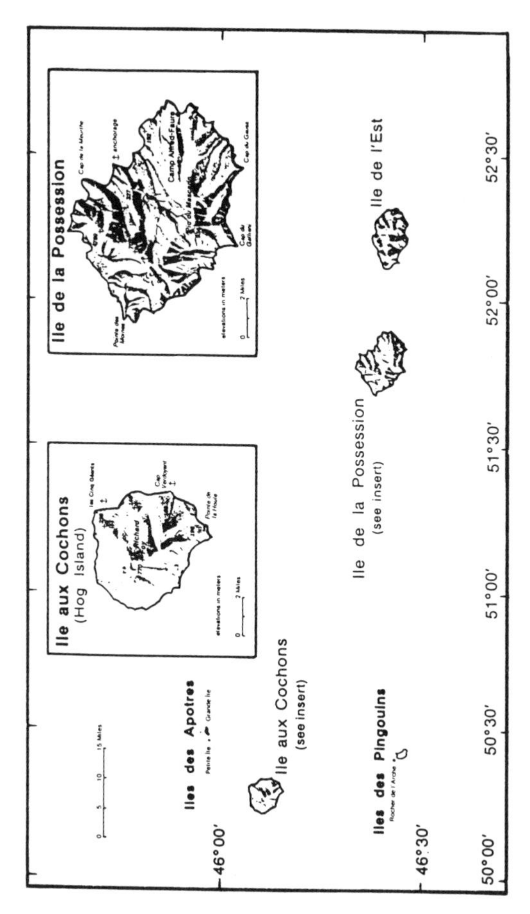

Crozet Islands (with Ile Aux Cochons and Ile de la Possession inset)

The Crozet Islands were discovered by a French expedition in 1772. Sealers and whalers were frequent visitors during the 19th century, but no settlement was attempted until the present-day scientific station was established. Ile aux Cochons, Hog Island, got its name when American whalers turned pigs loose to forage on the island so they could use them as a food supply to supplement their ships rations. The sea around these islands is full of treacherous reefs, and this, combined with the frequently stormy weather, has caused many shipwrecks. During a sealing voyage in 1819, the *Princess of Wales* was wrecked in the Crozets. Some of the crew managed to survive for 2 years until they were rescued and one of them, Charles Goodridge, wrote a book about his adventures. His *Narrative of a Voyage to the South Seas and the Shipwreck of the Princess of Wales Cutter with an Account of Two Years Residence on an Uninhabited Island* was published in 1843.

Today there is a research station, Camp Alfred-Faure, with a population of about 25 on Ile de la Possession. This camp began as a weather station that was set up in 1964. Temporary camps for special research have been set up from time to time on Ile aux Cochons. The research activities in the Crozets are closely tied to those at the larger French base at Port-aux-Francais in the Kerguelens. However, there is no transport between the two bases other than the two supply ships each year. Communication at other times is restricted to radio.

49. Archipel De Kerguelen

This French archipelago lies in the southern Indian Ocean at 48° to 50° south, 68° to 70° east. Kerguelen consists of over 300 islands, the largest of which is the very irregular, roughly triangular island Desolation. The whole archipelago has an area of 2,700 square miles. The coastline of the main island is deeply indented with many fiords, large bays dotted with islets and rocks, and many peninsulas. Inland, the island is cut up into numerous valleys and ridges. Mount Ross, the highest peak, reaches 6,068 feet. There is a permanent ice cap with glaciers flowing down from it. Ice covers one third of the island. The abundant rainfall and glacial meltwater keep numerous small rivers and lakes well supplied. The climate has only a small seasonal temperature variation with a summer average of around 40° F and a winter average of 37° F. Rain, snow or sleet occur on up to 300 days in a typical year, and the sky is usually overcast. Strong westerly winds prevail with gusts of up to 100 miles per hour. The winds retard the growth of vegetation and cause rain, snow, sleet and even hail to "fall" almost horizontally.

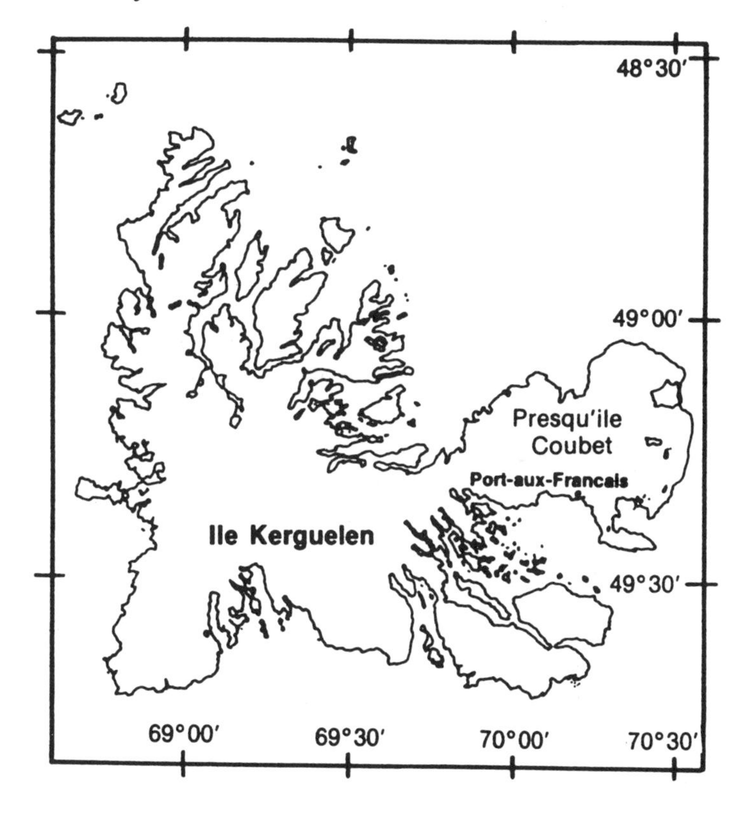

Ile Kerguelen

Resources found on the island include peat bogs, lignite and guano. Plant life is dominated by tussock grasses, mosses, lichen and the edible Kerguelen cabbage, which was formerly eaten by sealers to prevent scurvy. Offshore there are large patches of kelp. The Kerguelen cabbage is not only edible by man but also by the rabbits introduced by man. As the rabbits have increased in numbers, this once widespread plant has largely disappeared from those islands that are overrun by rabbits. Other animals brought to these islands include reindeer, hogs, sheep, dogs, cats, mice and rats. The reindeer and sheep have adapted to the rigors of island life. Cats thrive on Presqu'ile Courbet to the detriment of the bird population. But dogs were unable to survive in the wild. Mice have proliferated in the research base area, but rats have not. Along the shore one finds typical sub-Antarctic wildlife including penguins, seals, seabirds and ducks. Recently, salmon fingerlings were released in the inlets with the hope that they will survive and return to breed.

Kerguelen was discovered by a French navigator in 1772. In the 18th and 19th centuries it was visited frequently by sealing and whaling vessels, because Kerguelen's eastern bays offered the only protected anchorages in the turbulent seas of the southern Indian Ocean for ships seeking refuge from storms, shelter to make repairs, or replenishment of fresh water supplies. The anchorages were used again by marauding German warships in World War II. Early in the 20th century, an attempt was made at raising sheep on these islands, but that was abandoned in the 1930's. Now the French maintain a research station with a staff of about 90 people at Port-aux-Francais. Otherwise, the archipelago is uninhabited.

Although general aspects of these islands are known from aerial reconnaissance and from some coastal exploration by ship, much detailed exploration of these islands has not yet been done. The better known areas are on the peninsula named Presqu'ile Courbet on which Port-aux-Francais is located.

Port-aux-Francais is the main base for French research in the French Southern and Antarctic Lands. Scientists from other nations are occasionally invited to participate in the research programs. One of the more important of these is a joint French-Soviet effort in the fields of geomagnetism and ionospheric physics.

50. Amsterdam

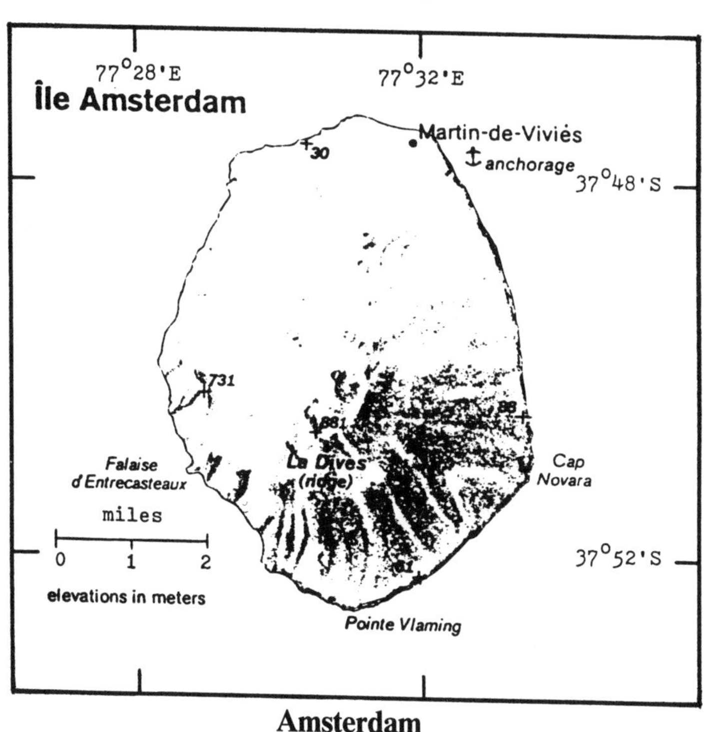

Amsterdam

This French island lies in the Indian Ocean 2,800 miles off the African coast and 1,200 miles northeast of Kerguelen at 38° south, 78° east. It is a moutainous island, 6 miles by 4 miles, with an area of about 21 square miles. Its highest elevation is 2,890 feet. Amsterdam, and the nearby island of Saint Paul, offer a more hospitable environment than the other islands of the southern Indian Ocean. Their latitude gives them a less rigorous climate in which horticulture is possible in protected spots. Both islands have an average summer temperature of about 61° F and an average winter temperature of 52° F. Very little snow falls in winter, and rainfall is moderate. In winter, strong winds and cloudy skies are the rule.

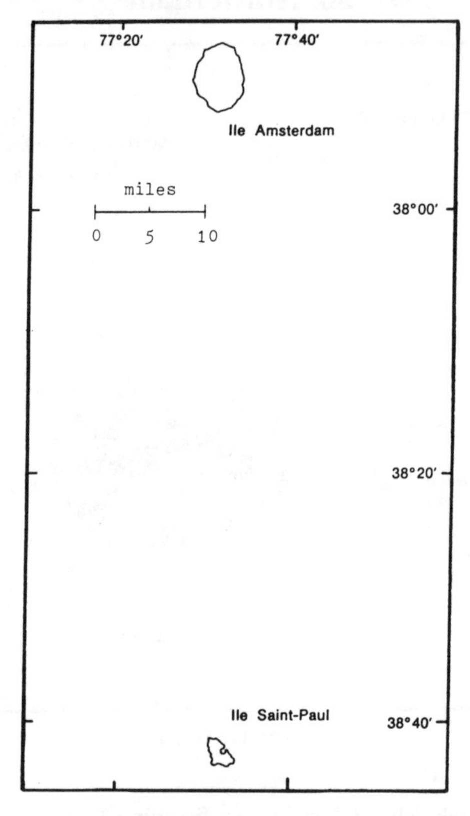

Ile Amsterdam and Ile Saint Paul

Amsterdam's broad north slope is dotted with small volcanic cones interspersed with bogs. A number of craters are filled with water. From a distance the surface appears to be open grassland, but the bogs are very difficult to traverse on foot.

Plant life is abundant. High stands of tussock grass are typical, and there are even a few wind-stunted trees. The animals found here were all introduced by man. They include cattle, cats, rats and mice. Some of the cattle brought to Amsterdam in the past survived in the wild state and multiplied, and now roam the island, but they have bone weaknesses, because of nutritional deficiences in the natural forage. Along the shore, seals and penguins breed. The surrounding waters are rich in marine life and have been fished commercially.

These islands have been known since the 16th century, and ships sailing between Europe and the East Indies often stopped here. Since the early 19th century, the waters around Amsterdam and Saint Paul have been fished by Reunion islanders at various times. In the mid 1830's, Reunionnais were taking a species of warm water cod from these waters and asked the French government to take possession of these islands to safeguard their interests in the fishing grounds. France declined, preferring instead to supply the Reunionnais with Newfoundland cod. But in 1893, France did assert its sovereignty over these islands, and today they are part of the French Southern and Antarctic Lands.

A weather station was installed on Amsterdam in 1950. This has grown until today it is the Martin-de-Vivies research station with a population of about 36. No other people live on the island. Since Amsterdam has no harbor, supplies for the research station are lightered to the base of a low cliff and lifted up from there by crane. Living conditions at the station are regarded as among the best in the southern islands — in part because of the warmer, more pleasant climate.

51. Saint Paul

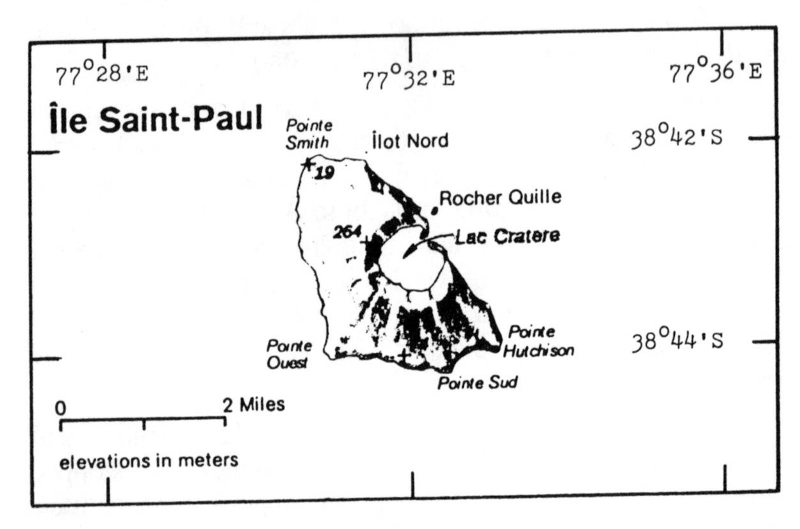

Saint Paul

Saint Paul lies about 60 miles south of Amsterdam at 39° south, 78° east. It is a smaller island, about 3 miles by 1½ miles, with an area of 3 square miles, and it consists of a broken volcanic crater which has been invaded by the sea. This submerged crater looks like a natural harbor, but it isn't used as a harbor because it is unprotected from gusty downdrafts that descend the crater slopes with little warning, and because access is limited by a depth of 7 feet across the mouth of the bay. Where Amsterdam has a high central peak, Saint Paul has a crater, so the smaller island only rises to 866 feet. Saint Paul has numerous hot springs. The island was the site of an unsuccessful settlement in the 18th century. Then again in the 20th century there was another attempt to exploit the island's resources that ended in tragedy. In 1928, a French company set up a seafood cannery on Saint Paul to process lobsters that are abundant near its shores. The company went bankrupt and apparently abandoned some 50 families who had been brought to Saint Paul as workers, mainly from Madagascar. A ship sent to check on the settlement 3 years later found only 2 survivors. Currently, 2 vessels from Reunion visit the waters off Saint Paul each year to harvest the quota of 350 tons of lobster tails.

52. Heard Island

About 300 miles southeast of Kerguelen at 53° south, 74° east, one finds Heard, which consists of a circular mountain, known as Big Ben, that rises from the ocean, with a smaller peak attached to one side. The summit of Big Ben, a volcanic cone named Mawson Peak, rises 9,004 feet, and the lower Anzac Peak on the Laurens Peninsula rises to 2,345 feet. A permanent ice cap covers most of the island ending in sheer cliffs along much of the coast. Heard occupies an area of 146 square miles.

The climate of Heard is harsh, due to its location in the "freezing fifties" of latitude. Westerly winds prevail throughout the year, and severe storms, accompanied by high winds and rain and snow may strike with little or no warning. Periods of good weather — lasting occasionally as long as 2 weeks — are most likely to occur in December and January. Snow covers the entire island from April through November. The average temperature the year round is only 33° F. These environmental discomforts — stormy seas, howling gales and somber, overcast skies — magnified on this lonely outpost, make Heard the hardship post among the research stations of the southern Indian Ocean.

The heavily glaciated terrain is desolate and treeless; vegetation is limited to lichen and mosses and to small flowering plants in season. Seals, and penguins and numerous other bird species, are found on the island. Although wanton slaughter in the 19th century greatly thinned the seal and penguin populations, recent efforts to conserve these animals have been successful. A 1969 census showed a considerable increase over the 1963 count.

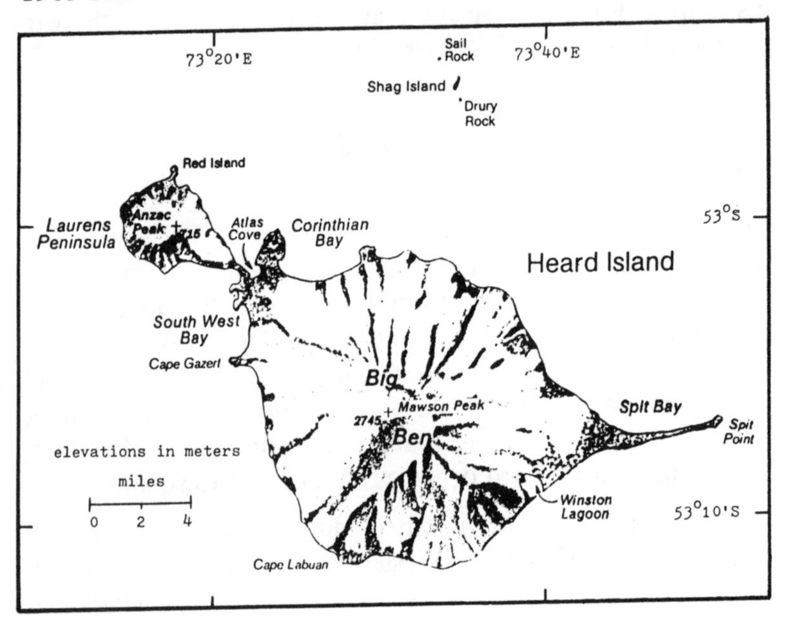

Heard

Heard Island was discovered by a British sealer, Peter Kemp, in 1833, but the first landing was made by an Australian, Captain Heard, in 1855. Fur seal hunters from the US later visited regularly, dominating the seal hunting until 1890. Great Britain laid claim to the island in 1910 and retained control until 1947, when administration was transferred to Australia.

Australian reconnaissance parties have made sporadic visits to study ionospheric physics, glaciation, volcanic activity and plant and animal life. The Australian National Antarctic Expedition operated a weather station on Heard from 1947 until their Mawson Station was established in Antarctica in 1955. Weather observations are now limited to infrequent Australian summer visits. Reportedly, permission has been granted by the Australians to the Soviet Union to station a research group on the island.

Poor weather conditions have prevented detailed study of Big Ben. Two members of a 1963 expedition reached the main crater but were forced to return to camp after 5 days of bad weather without being able to measure the activity of the volcano. Steam has been observed emerging from the crater and from fumaroles on the side of the cone.

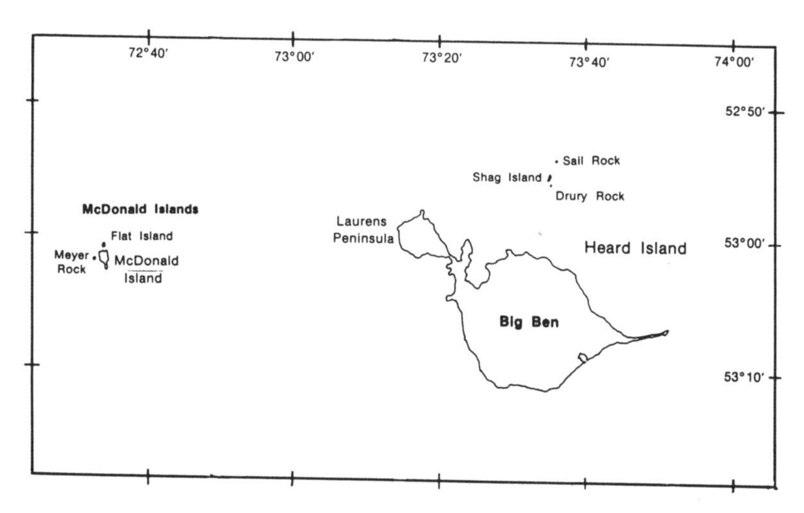

Heard and McDonald Islands

53. McDonald Islands

The McDonald Islands are a group of 3 small, uninhabited islets — McDonald Island, Flat Island and Meyer Rock — about 30 miles west of Heard Island at 53° south, 73° east. The first known landing on McDonald, largest of the 3, was made by helicopter in 1971 when two members of a French/ Australian survey team spent 45 minutes reconnoitering the island. They verified that South and Needle, which were formerly thought to be separate islands, are in reality promontories of McDonald. Elevations range from about 400 feet in the north, to more than 650 in the south. All 3 islands are treeless and, like Heard, are used as breeding grounds by various seabirds, penguins and seals. Along with Heard, the McDonald Islands have belonged to Australia since 1947.

Part V
Other Islands

No single book could cover all of the many thousands of uninhabited islands around the world. Here we'll briefly mention some other deserted islands not discussed in this book.

a. There are more than 1,000 islands in the Adriatic Sea, off the coast of Yugoslavia, and only 62 of them are populated.

b. The Orkney Islands, north of Scotland, consist of 90 islands, only 29 of which are inhabited.

c. There are 700 islands in the Bahamas, east of Florida, and only 15 of them are inhabited.

d. Of the 7,000 islands in the Philippines, about 6,000 are uninhabited.

e. There are hundreds of islands off the coast of New Zealand that are not inhabited.

f. The coasts of British Columbia (western Canada) and Alaska are a maze of islands and many are not inhabited.

g. There are possibly thousands of islands off the coast of Australia that are uninhabited. Most of the northwest coast between Port Headland and Darwin has not been charted. Admiralty charts of this area say: "numerous islands and reefs — not charted."

References

Columbia Lippincott Gazetteer of the World, 1952. An alphabetical listing of place names, including islands, which gives a brief description of each entry.

Danielsson, Bengt, *Forgotten Islands of the South Seas,* 1957. An account of the year he spent in the Marquesas. This is a somewhat unfavorable report which seems to emphasize the degeneracy and diseases of the native populations.

Eliot, John L., "Hawaii's Far-Flug Wildlife Paradise," in *National Geographic,* May 1978. This article includes aerial photos of Laysan, Kaula and Whale-Skate Island (French Frigate Shoals).

Encyclopedia Americana, 1978. The entry "Pacific Ocean" has a long list of islands with their populations, indicating which islands are uninhabited.

Fisher, Jon, *The Last Frontiers on Earth,* published by Loompanics Unlimited, 1980. This book gives information on how to live in polar regions and on or underwater, as briefly mentioned in the Introduction.

Furse, Chris, *Elephant Island, an Antarctic Expedition,* 1979. Relates the experiences of a December 1976 to March

1977 British scientific survey of the Elephant Island group, which mapped these islands accurately for the first time.

Handy, Willowdean C., *Forever the Land of Men,* 1965. A traveler's account of a visit to the Marquesas in 1920 and 1921 with an ethnological expedition. The people and all the uninhabited islands are described, and the book includes small maps of all the inhabited islands.

Henderson, et. al., *Area Handbook for Oceania,* 1971. Has information on the Marianas, Carolines and Marshalls in an appendix.

Heyerdahl, Thor, *Fatu Hiva, Back to Nature,* 1974. Tells about the year (1936) Thor, then 22, and his wife Liv spent living off the land on the southernmost Marquesas island of Fatu Hiva. He describes the climate, vegetation and physical conditions on Fatu Hiva, which are similar to the other Marquesas. He also tells a little about Hiva Oa, and says all that needs to be said about uninhabited Motane.

Indian Ocean Atlas, by the CIA, August, 1976. Superintendent of Documents, US Government Printing Office, Washington, DC 20402, stock number 041-015-00080-2. This is a thorough survey of the environment, resources, shipping and politics of all Indian Ocean islands, complete with dozens of excellent maps.

Irvine, Lucy, *Castaway,* 1983. Describes the year (May 1981 to June 1982) that Lucy Irvine and Gerald Kingsland spent on the uninhabited one mile long island of Tuin in the Torres Strait north of Australia. They set out to live off the land with meager supplies and were saved from starvation by the timely assistance of visitors from a nearby island. Paints a vivid picture of life on a small, remote tropical island.

Islands of the Pacific. An informative 23 by 37 inch map available from National Geographic Society, PO Box 2806, Washington, DC 20013.

Kahn, E.J., Jr., *A Reporter in Micronesia,* 1966. Describes the Trust Territory of the Paficic Islands (Mariannas, Palau, Carolines and Marshalls).

King, H.G.R., *The Antarctic,* Arco Books, New York, 1969. A thorough discussion of the physical conditions, geography and history of the Antarctic region, including a chapter describing in detail each of the sub-Antarctic islands.

Maps of Islands. Very fine maps of many islands are sold by two US Government agencies:

1. Defense Mapping Agency, Office of Distribution Services, Washington, DC 20315.

2. National Ocean Survey, US Department of Commerce, NOAA/NOS — C44, Riverdale, MD 20737.

May, John, *The Greenpeace Book of Antarctica,* 1988. A survey of Antarctica from Greenpeace's ecological perspective. Sub-Antarctic islands are mentioned in connection with discussion of the scientific stations located on them.

McLintock, A.H., editor, *An Encyclopedia of New Zealand.* This 3-volume reference work, published by the Government of New Zealand, contains excellent descriptions of all the sub-Antarctic islands lying near New Zealand.

Neale, Tom, *An Island to Myself.* Neale lived alone for several years on Suvarov, an atoll in the Cook Islands, and his book tells about his life there. He gives a detailed account of conditions on this typical atoll and reveals what it is like to live alone on one.

Norgrove, Ross, *Blueprint For Paradise, How to Live on a Tropical Island,* 1983. Covers all necessary details for anyone considering moving to a tropical island of whatever type, from a remote, uninhabited type 1 island, to a type 5, populated island with a town, utilities, and all amenities.

Polar Regions Atlas, by the CIA. From US Government Printing Office, stock number 041-015-00094-2. An excellent, detailed survey of the Arctic and Antarctic with dozens of maps.

Robson, R.W., *Pacific Islands Yearbook.* This is a standard reference for Pacific Islands, full of detailed information, prepared by the publisher of the Australian magazine *Pacific Islands Monthly.* The 13th edition was published in 1978. New editions of the *Yearbook* are published every few years and can be obtained from: Pacific Publications Pty., Ltd., Technipress House, 29 Alberta St., Sydney, NSW 2000, Australia.

Silverman, David, *Pitcairn Island,* 1967. A thorough description of the history, and the physical and social situation on Pitcairn. There are also some brief mentions of Oeno, Henderson and Ducie. Pitcairn has a tiny population, but it is of interest because it may become depopulated. Also, Pitcairn is similar to the uninhabited Kermadec Islands, which are situated at about the same latitude, although thousands of miles away across the Pacific. And, anyone thinking about establishing a small community on an empty island would be well advised to study Pitcairn to get a taste of the social situation which is likely to result.

Skaggs, Jimmy M., *Clipperton, A History of the Island the World Forgot,* 1989. This 300 page book reveals about all there is to know about this obscure island. The author has dug up just about everything that is known to have been written about Clipperton, and he describes every known visit to the island.

Stanley, David, *South Pacific Handbook,* 3rd edition, 1986. Moon Publications, PO Box 1696, Chico, CA 95927. This is a very thorough traveler's guide to the South Pacific Islands, featuring descriptions of major islands and cities, and information such as where to stay and what to see. There are passing references to a few uninhabited islands.

Stanley, David, *Micronesia Handbook,* 1985. Another in the Moon Publications series of guidebooks (see address above), with some brief mentions of a few uninhabitaed islands in Micronesia.

Strange, Ian J., *The Falkland Islands,* 1972. Stackpole Books, Harrisburg, PA 17105. Covers the physical geography, history and pre-1982 war political situation in the Falklands. This is a good introduction which is lacking in only one area: it gives no hint of the feudalistic land holding system, the absentee landlordism, and the strong class feeling that dominates the social structure on these islands.

The Times World Index Gazetter, 1965. Published by the *London Times.* Lists the latitude and longitude of 345,000 geographical places, including most islands.

Webster's New Geographical Dictionary, published by the G & C Merriam Co., Springfield, MA, 1972. A dictionary of geographical places around the world, including islands, with a brief description of each.

Index

(Name of Island, followed by page number)

List of Maps (Alphabetical, with page number)